SpringerBriefs in Applied Sciences and Technology

Thermal Engineering and Applied Science

Series editor

Francis A. Kulacki, University of Minnesota, Minneapolis, MN, USA

SpringerBriefs present concise summaries of cutting-edge research and practical applications across a wide spectrum of fields. Featuring compact volumes of 50 to 125 pages, the series covers a range of content from professional to academic.

Typical publications can be:

- A timely report of state-of-the art methods
- An introduction to or a manual for the application of mathematical or computer techniques
- A bridge between new research results, as published in journal articles
- A snapshot of a hot or emerging topic
- An in-depth case study
- A presentation of core concepts that students must understand in order to make independent contributions

SpringerBriefs are characterized by fast, global electronic dissemination, standard publishing contracts, standardized manuscript preparation and formatting guidelines, and expedited production schedules.

On the one hand, **SpringerBriefs in Applied Sciences and Technology** are devoted to the publication of fundamentals and applications within the different classical engineering disciplines as well as in interdisciplinary fields that recently emerged between these areas. On the other hand, as the boundary separating fundamental research and applied technology is more and more dissolving, this series is particularly open to trans-disciplinary topics between fundamental science and engineering.

Indexed by EI-Compendex, SCOPUS and Springerlink.

More information about this series at http://www.springer.com/series/8884

Yang Liu • Francis A. Kulacki

The Effect of Surface Wettability on the Defrost Process

 Springer

Yang Liu
Graduate School at Shenzhen
Tsinghua University
Shenzchen, China

Francis A. Kulacki
Department of Mechanical Engineering
University of Minnesota
Minneapolis, MN, USA

ISSN 2191-530X   ISSN 2191-5318   (electronic)
SpringerBriefs in Applied Sciences and Technology
ISSN 2193-2530   ISSN 2193-2549   (electronic)
SpringerBriefs in Thermal Engineering and Applied Science
ISBN 978-3-030-02615-8   ISBN 978-3-030-02616-5   (eBook)
https://doi.org/10.1007/978-3-030-02616-5

Library of Congress Control Number: 2018959755

This Springer imprint is published by the registered company Springer Nature Switzerland AG
The registered company address is: Gewerbestrasse 11, 6330 Cham, Switzerland

Preface

Frost formation and anti-icing methods have been investigated for decades. In refrigeration systems, a number of studies were focused on the influence of surface treatment in delaying the time of frost formation. There are far fewer studies on the effect of surface wettability on the defrost process. The literature contains experimental results on defrost time for different surfaces, but the results can be different based on individual experimental conditions.

In this monograph, the influence of surface wettability on the defrost process is investigated analytically and experimentally. The melting process is divided into three stages based on behaviors of the meltwater. Water saturation and meltwater draining velocity are formulated for the absorption and drainage stages separately. The melting rate, permeation rate, and draining velocity determine meltwater behaviors, which influence the defrost process and defrost mechanisms. Water accumulation at the surface decreases the adherence of the frost column to the surface, and thus, frost slumping is a potential method for frost removal. A slumping criterion is formulated based on the analysis of interfacial forces and the body force on the frost column. The slumping condition of the model depends on the contact angle on hydrophilic surfaces and the contact angle hysteresis on hydrophobic surfaces.

Experiments on frost and defrost process on vertical surfaces with different wetting conditions are conducted in a laboratory environment. The experiments show that defrost time and efficiency are determined by system design, surface heating method, heat flux applied at the surface, and surface wettability. Defrost mechanisms vary with surface wettability. During the defrost process, the frost column detaches from a superhydrophobic surface and falls off as a whole piece. While on the superhydrophilic and plain surfaces, the frost column melts, and the water film or retained water evaporates. Defrost time and efficiency are not significantly different on the test surfaces at the point that the frost melts and water film or retained droplets remain on the surface. However, defrost time and efficiency are improved noticeably on the superhydrophobic surface for a complete defrost process in which the water evaporation time is included.

This research serves as a supplement to current studies on the effect of surface wettability in refrigeration systems. In application, the choice of surface wettability factors depends on practical operation requirements and system design.

Shenzchen, China   Yang Liu
Minneapolis, MN, USA   Francis A. Kulacki

Contents

Nomenclature

b	Slip length (m)
bl	Gap between fins (m)
c_p	Specific heat at constant pressure (J kg^{-1} K^{-1})
g	Gravitational acceleration (m s^{-2})
h	Heat transfer coefficient (W m^{-2} K^{-1})
k	Thermal conductivity (W m^{-1} K^{-1})
m	Mass (kg)
m''	Mass flux (kg s^{-1} m^{-2})
p	Permeability power law constant
q	Capillary pressure power law constant
q''	Heat flux (W m^{-2})
r	Aspect ratio of the width to the length of the test surface
t	Time (s)
t^+	Dimensionless time
u	Velocity in x-direction (m s^{-1})
u^+	Dimensionless velocity
u_s	Slip velocity (m s^{-1})
u_B	Velocity at the water–permeation interface (m s^{-1})
v	Velocity in y-direction (m s^{-1})
x	Direction parallel to gravity (m)
x^+	Dimensionless location
y	Direction normal to the test surface (m)
v_b	Bulk flow rate (m s^{-1})
v_s	Slip velocity (m s^{-1})
A	Area (m^2)
Bi	Biot number, hL_c/k
C	Entry capillary pressure (N m^{-2})
CAH	Contact angle hysteresis (°)
CFM	Cubic feet per minute

D	Diameter (m)
F	Force (N)
FR	Force ratio of gravity to surface tension
Fr	Froude number, $u/(gl)^{1/2}$
G	Gravity (N)
Gr	Grashof number
H	Height (m)
K	Permeability or intrinsic permeability (m^2)
L	Length (m)
L_f	Latent heat of fusion (kJ kg^{-1})
Nu	Nusselt number
P	Pressure (Pa)
P^+	Dimensionless pressure
P_{ca}	Capillary pressure
Pr	Prandtl number
Pwr	Power (W)
Q	Heat transfer (J)
T	Temperature (K)
U	Characteristic velocity (m s^{-1})
V	Volume (m^3)
W	Width (m)
Re	Reynolds number
R_{th}	Thermal resistance (W K^{-1})
S	Water saturation, the volume fraction of water to the pore volume

Greek Symbols

α	Thermal diffusivity (m^2 s^{-1})
β	Dimensionless parameter depending on the material
δ	Thickness (m)
ε	Porosity
η	Efficiency
θ	Contact angle ($^\circ$)
Θ	Dimensionless temperature
ϕ	Slope angle ($^\circ$)
φ	Relative humidity
μ	Dynamic viscosity (N s m^{-2})
μ_s	Average viscosity of the near-to-wall layer (N s m^{-2})
ρ	Density (kg m^{-3})
γ	Surface tension (J m^{-2} or N m^{-1})

τ	Shear stress ($\mathrm{N\ m^{-2}}$)
ω	Filtering velocity of water in the permeation layer ($\mathrm{m\ s^{-1}}$)
Ω	Volumetric flow rate ($\mathrm{m^3\ s^{-1}}$)

Subscripts

0	Initial
a	Air
adv	Advancing
Al	Aluminum
avg	Average
b	Bulk
c	Cold
ca	Capillary
ch	Chamber
cond	Conduction
conv	Convection
d	Drain
df	Defrost
eff	Effective
f	Frost
g	Gas
h	Hot
i	Ice
in	Input
kp	Key point
lat	Latent
m	Melt
p	Permeation
rec	Receding
s	Surface
sat	Saturation
sen	Sensible
s-p	Solid–permeation interface
s-w	Solid–water interface
tot	Total
tp	Test plate
v	Vapor
w	Water
w-p	Interface at water film and permeation layer

Chapter 1
Introduction

Frost formation is a common problem across a wide range of technologies. It can prevent normal operation in refrigeration systems, power lines, wind turbines, and aircraft. In refrigeration systems, frost formation on evaporator surfaces can block airflow, increase thermal resistance and pressure drop, and decrease the coefficient of performance. Consequently, a defrost process is necessary to maintain normal operation.

The defrost process involves heating the frost column, frost melting, and evaporation of residual melt water. Frost melting is accompanied by water draining, but the physical mechanisms of the melting process have not been thoroughly investigated to cover the different meltwater behaviors. Recently passive deicing methods have attracted interest as a way to speed the defrost process and thus reduce down time for the refrigeration system. The use of hydrophilic and hydrophobic surfaces is thought to influence frost growth and the defrost process. Experimental results show that a hydrophobic surface can retard the frost process during nucleation and at the beginning of frost growth. However after the surface is covered with frost, the effect of surface wettability shows no significant difference in the frost growth period.

The current literature on the effect of surface wettability on the defrost process is limited. Compared to hydrophobic surfaces, a hydrophilic surface shows much less water retention after melting as the water film drains more easily from it. The conditions under which experiments have been performed differ with a range of surface temperatures, air temperatures, and ambient humidity values. Also only few experimental results are available on the effect of surface wettability on the defrost process. It is difficult to conclude whether hydrophilic surfaces are superior to hydrophobic surfaces or vice versa.

The theoretical study of the defrost process is focused on defrost models. Many researchers have investigated defrost models for electrical defrosting and hot gas defrosting. However the effect of surface wettability has not been fully discussed in current models. Another observation from the current literature is that frost slumping, which is the behavior of the bulk frost mass peeling (or dropping) off

© The Author(s), under exclusive license to Springer Nature Switzerland AG 2019 1
Y. Liu, F. A. Kulacki, *The Effect of Surface Wettability on the Defrost Process*,
SpringerBriefs in Applied Sciences and Technology,
https://doi.org/10.1007/978-3-030-02616-5_1

the surface, has not been analyzed theoretically and experimentally. Slumping is a favorable phenomenon during the melting process. If the bulk mass releases from the surface, defrost time reduces significantly, and the defrost efficiency would improve greatly with less energy consumption.

In this monograph we develop an analytical model of the defrost process with an emphasis on the effect of surface wettability. The effect of surface wettability is expected to be significant during the melting period and to have an effect on the initiation of slumping. The defrost model is governed by coupled heat and mass transfer within the frost layer, and the slumping model is dominated by the balance of surface tension, adhesion, and gravity. However, the two models are connected in nature. Surface tension and ice adhesion are functions of temperature which can be obtained from the defrost model. Specific objectives are to formulate the analytical model to describe the melting process with a good understanding of the physical mechanisms, study the effect of surface wettability on the defrost process, investigate the criterion when slumping might happen, and test the defrost process on the different surface wettability. The effect of surface wettability on the defrost process is a new feature of this research and has theoretical and practical importance. We provide a mathematical model that relates defrost time and efficiency to the surface wettability, including a range of meltwater behaviors. The slumping model is expected to provide a new prospect on frost removal methods and might be applied to design and optimization of refrigeration systems.

Chapter 2
Prior Research

2.1 Frost Formation

The defrost process initiates with a frost layer of a certain thickness and bulk porosity. Frost formation involves nucleation, crystal growth, and frost layer growth. In this review of prior research [1–26], we focus on frost nucleation, empirical models and correlations, and analytical models to predict frost growth.

Frost Nucleation

Frost forms by either condensation of water vapor or freezing of liquid water. The initiation of frost formation involves a nucleation process, which is categorized as either homogeneous or heterogeneous nucleation. Homogeneous nucleation is also called spontaneous nucleation with water molecules combining together to form a frost embryo. Heterogeneous nucleation occurs on foreign particles and is more common in nature. The discussion herein refers to heterogeneous nucleation.

Frost nucleation may be affected by water vapor pressure, surface temperature, and surface energy. Na and Webb [1] analyze the thermodynamic process of nucleation and conclude that supersaturation of water vapor is a requirement for frost nucleation. They point out that the degree of supersaturation[1] depends on surface energy.[2] Lower energy surfaces require higher supersaturation for nucleation than higher energy surfaces, and this analysis agrees with the experimental results. They test five surfaces with different surface energies, or contact angles: hot water treated aluminum surface, bare aluminum surface, transparent polymer packaging

---

[1]Degree of super saturation is the humidity ratio difference of the air stream and the surface.

[2]Surface energy is the sum of all the excess energies of atoms at the surface.

© The Author(s), under exclusive license to Springer Nature Switzerland AG 2019
Y. Liu, F. A. Kulacki, *The Effect of Surface Wettability on the Defrost Process*,
SpringerBriefs in Applied Sciences and Technology,
https://doi.org/10.1007/978-3-030-02616-5_2

tape on the aluminum surface, silicone wax-coated aluminum surface, and Teflon-coated aluminum surface. Experiments show that frost formation starts at discrete points, and some regions are not covered by frost at a certain temperature drop. This inhomogeneity results from a variable surface energy. They observe that surface roughness reduces the required supersaturation for nucleation as the contact area between the nucleation embryo and the substrate increases. Their results imply the possibility of surfaces that may delay frost formation.

The influence of contact angle on supercooling[3] is also reported by Piucco et al. [2]. They propose a nucleation model based on classical nucleation theory, in which the minimum energy barrier of nucleation is expressed as a function of the degree of supersaturation and contact angle. Their model applied to smooth surfaces.[4] They also show that nucleation limits are a function of the supercooling and contact angle and are independent of surface temperature. Their experimental work was carried out in an open loop tunnel, and five surfaces, each with a different surface treatment, were tested under a fixed environmental condition. Results show that greater surface roughness favors the occurrence of nucleation. The experimental results show agreement with the theoretical prediction.

Experiments and Correlations

A number of experimental studies [3–10] have been done to investigate thermophysical properties and empirical correlations during frost growth. Quantities of interest include density, thermal conductivity, thickness, mass, growth rate, and heat and mass transfer coefficients. Both environmental and surface conditions affect the properties. Environmental conditions include air temperature, velocity, and humidity. Surface conditions include surface temperature and treatment.

Thibaut Brian et al. [3] investigate frost densities, thermal conductivities, and heat and mass fluxes as a function of time under the conditions of ambient temperature of 1.1–33.9 °C, dew point of -10–14.4 °C and Reynolds number of 3770–15,800. Measurements agree with predictions of a simple analytical model emphasizing internal diffusion within the frost layer. The experimental results show no significant density gradients within the frost but are not thoroughly explained. It is postulated that very small ice particles are nucleated in the frost, and the nuclei are transported by thermal diffusion forces.

Hayashi et al. [5] relate frost properties to frost formation types. Their experiments are set under forced convection with ambient temperature from 15 to 30 °C. Frost formation is divided into three periods: crystal growth, frost layer growth, and

---

[3]Supercooling is defined as the temperature difference between the surface temperature and ambient saturation temperature.

[4]Smooth surfaces are surfaces for which the deviations in the direction of the normal vector of a real surface from its ideal form are small.

frost layer full growth. The crystal growth period is characterized by crystal growth in vertical direction, and the frost shape is like a forest of trees. During the frost layer growth period, frost grows into a meshed and uniform frost layer. The frost layer full growth period is characterized by melting, freezing, and deposition, and the frost shape does not change. The frost layer becomes denser in this period. Frost formation types are then classified into four groups based on structure which is determined by humidity and the cold surface temperature. Experiments show that frost density increases rapidly at the early stage. A higher ambient velocity produces a denser frost, but frost formation types are almost not affected by the ambient velocity. Thermal conductivity is shown to be not only a function of density but also a function of other factors. The structure of the frost layer, internal diffusion of water vapor, and surface roughness may also affect thermal conductivity.

An experimental study of frost properties under free convection is described by Fossa and Tanda [7]. Frost thickness, frost surface temperature, deposited mass and heat flux are investigated on a vertical plate. The plate is placed in a vertical channel with air temperature of from 26–28 °C, ambient relative humidity of 31–58% and test surface temperatures of -13 to -4 °C. The frost layer is not uniform during the early time period, and the thickest frost occurs on the coldest plate and at the highest relative humidity. The frost surface temperature is close to the triple-point value at high humidity. Deposited frost mass increases linearly with time and is more affected by the relative humidity than by the surface temperature. Heat flux at the plate–frost interface is measured and is shown to be a function of relative humidity and plate temperature. Local heat flux decreases markedly on the coldest plate. Total heat flux is higher at the higher relative humidity, which is explained by the increased rate of latent heat transfer.

Janssen et al. [10] present a new correlation and model based on digital reduction of in plane observations of frost thickness. Visual and digital methods are employed to determine frost thickness, porosity, and droplet statistics. Experiments show frost thickness increases with time exponentially, and a fast and slow growth period are identified. Frost thickness is expressed as a function of system state and growth resistance. The key physical and correlating parameter within this model is the ratio of sensible heat transfer to total heat transfer, and the growth rate varies inversely with this ratio. The resulting correlation faithfully captures measured growth rates and gives better predictive capability than that of existing correlations across a wide spectrum of frosting conditions. Experimental uncertainty on frost thickness is reduced of 0.1 mm, and the need is eliminated for measurement of the frost–air interface temperature. Janssen et al. also present a comprehensive review of experimental studies of frost growth illustrating the wide variety of techniques and frosting conditions employed over the past 30 years.

Analytical Models

Analytical study of frost formation is focused on models to predict frost densification and thickness, thermal conductivity, and growth rate [11–24].

Tao et al. [14] present a one-dimensional theoretical model to predict temperature distribution, frost density and thickness. The model involves two stages: ice column growth and fully developed frost layer growth. The frost layer is treated as a porous medium during the frost growth period. Both temporal and spatial variations of frost properties are analyzed using the local volume averaging technique. Calculations show that ice column diameter at the frost surface is smaller than that at the cold plate surface during the ice column growth period. During the frost layer growth period, temperature and vapor density show nonlinear distribution within the frost layer. The rate of densification has a maximum absolute value near the frost–air interface, the warm side of the frost layer. The experimental study shows good agreement with the theoretical calculations. A two-dimensional model is analyzed by Tao and Besant [15]. The initial stage is modeled with convective heat and mass transfer over the ice columns rather than diffusion within the frost. After a transition time, a homogeneous porous medium model applies to the frost layer. Vapor diffusion is considered to be the mechanism for water vapor transport within the frost layer, and phase change is due to sublimation. Calculations show good agreement with the reported experimental data.

Lee et al. [19] present a model to predict the thickness, density, and surface temperature within an error band of 10% during the frost growth period. Assumptions include that all the processes are quasi-steady state, and the variation of frost density in the normal direction is negligible. Governing equations on the air side and in the frost layer are solved numerically. Experiments show that the frost layer thickness increases rapidly in the early stages if growth and then slows down. Frost density also increases rapidly at the beginning and then slows down. The model predicts the measured results accurately except during the early stage of the frost formation, which may be attributed to the experimental uncertainty. The surface temperature increases rapidly at beginning but the increase rate slows down with the decrease of the growth rate. The heat transfer rate decreases rapidly at the early stage but the rate of decrease slows with time. The trend in the mass transfer rate is similar to that of the heat transfer rate.

Compared to research on frost growth, theoretical studies of the crystal growth are limited. Sahin [25, 26] investigates frost properties and presents an analytical study of the crystal growth period. The frost layer is assumed to consist of several frost columns. The crystal structure varies with temperature. The temperature distribution shows that more sublimation of water vapor occurs in the upper half of the frost layer. The temperature distribution in the frost layer is mostly affected by the ambient humidity, ambient temperature, and surface temperature. The Reynolds number has less effect on the temperature distribution. The effective thermal conductivity is found to be not only a function of density but also depends on vapor diffusion and sublimation. The effective thermal conductivity increases rapidly with time as densification occurs and levels off as the frost layer thickens. Higher surface temperature and higher air temperature yield larger effective thermal conductivity. Low humidity results in nearly constant thermal conductivity, which is explained by the negligible vapor diffusion in the frost layer. Reynolds numbers have less effect on the frost thermal conductivity.

Effects of Surface Wettability

The effect of surface wettability on frost formation has attracted many researchers in the last decade. Studies are mainly focused on the experimental observations of frost properties corresponding to surface wettability and frost formation with respect to surface wettability. Some analytical models have been proposed with the effect of surface wettability included during frost formation. A number of investigations [28–38] address anti-icing performance of the superhydrophobic surfaces.

Shin et al. [27] investigate the effects of surface energy on frost formation and provide correlations for frost layer properties as a function of time and surface energy. The tests are performed on three different surfaces with dynamic contact angles[5] (DCAs) of 23°, 55°, and 88°. Experimental conditions are maintained at an external Reynolds number of 9000, relative humidity 48.1%, air temperature 12 °C, and cold plate temperature −22 °C. The results show that frost thickness increases rapidly with time at the beginning of frost growth and then approaches a constant rate of growth. Frost thickness on the surface with a 23° DCA is 76% of that with 88° DCA after 30 min, but the difference is not significant after 120 min. Filmwise condensation and dense droplets form on lower dynamic contact angle surfaces. Dropwise and less dense condensate are generated on higher DCA surfaces. Frost mass increases linearly with time, and the differences on tested surfaces are minor. Frost density is larger on lower DCA surfaces at the initial stage, and the difference becomes less significant after 120 min. Thermal conductivity on surfaces with different DCAs shows smaller variation than density. Local thermal conductivity varies significantly along the frost layer and conductivity close to the surface and the frost–air interface is two to three times larger than that inside the frost layer. Zhong et al. [28] present a method of producing superhydrophobic surfaces with micro-posts and micro-grooves. Water droplets form on the surface with static contact angle (SCA)[6] of 148°. Experiments demonstrate that condensate appears everywhere when droplets are smaller than the characteristic dimension of the microstructure. As condensate accumulates, droplets either group around the posts or condense and form a thin film depending on the micro-structure. This phenomenon influences

---

[5]Dynamic contact angle is measured when the three phase boundary is moving. Dynamic contact angles are referred to as either a advancing or receding contact angle.

[6]Static contact angles (SCA) are measured when a droplet is standing on the surface, and the three phase boundary is not moving. Contact angle in this monograph is referred to as static contact angle.

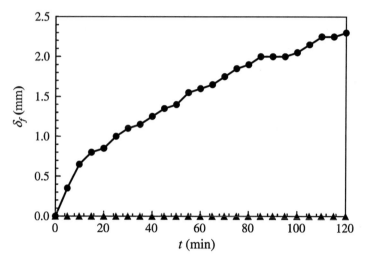

Fig. 2.1 Comparison of frost deposition on coated and uncoated surfaces [29]. $T_a = 15.4$ °C, $\phi = 0.57$, $T_s = -8.2$ °C. ▲—coated surface. ●—uncoated surface

condensate drainage and frost formation. The density and thermal conductivity of the frost layer vary significantly for frost growing from dropwise condensate compared to frost growing from filmwise condensate.

A long delay time is reported by Liu et al. [29] for frost formation for surfaces coated with an anti-frosting paint. They show that surface temperature and ambient humidity exert a strong influence on the performance of the coating. Their experiments show that a no frost condition is maintained for up to 3 h (Fig. 2.1.) when $\phi < 0.6$ and $T_s > -10$ °C. The influence of hydrophobicity on frost formation on a vertical plate under natural convection conditions is also studied by Liu et al. [30]. A hydrophobic surface with an SCA $= 133°$ and plain copper surface with an SCA $= 63°$ were tested, and experiments suggest that smaller droplets are formed on hydrophobic surfaces. The explanation is that the contact area on the hydrophobic surface is smaller than that on a hydrophilic surface if the volumes are equal. It is also observed that frost crystal growth on the hydrophobic surface has a stronger dendrite pattern resulting into looser and weaker frost layer at the initial stage. Frost thickness on the hydrophobic surface is smaller than that for a hydrophilic surface at the beginning of frost growth but approaches that for the hydrophilic surface when frost fully covers the surface. The potential barrier[7] relation shows that increasing contact angle increases the potential barrier and thus retards the frost crystal nucleation in early stage frost formation. After the surface is covered by frost, frost thickness growth on hydrophobic and hydrophilic surfaces is almost the same. Liu et al. [31] further examine frost formation on a superhydrophobic surface under

---

[7]The potential barrier is the Gibbs free energy difference when supersaturated vapor changes from its metastable phase into the stable phase (frost crystal).

natural convection. Their experiments are performed on a horizontal surface with droplet SCA = 162°, $\phi = 0.4$, $T_s = -10.1$ °C, and $T_a = 18.4$ °C. Results show that frost formation is delayed for 55 min compared to the plain copper surface, and the frost layer structure on the superhydrophobic surface is looser (less dense) and easily removed. A frost layer pattern similar to a cluster of chrysanthemum petals is firstly reported in their paper. Water droplets on the superhydrophobic surface grow in a similar way as on the normal copper surface, but frost crystals grow in a different way. Frost crystals grow horizontally (along the surface) compared to the vertical direction on normal copper surface.

Some studies show the retarding effect of the superhydrophobic surface on the frost formation at the early stage. Wang et al. [33] generate superhydrophobic surfaces with contact angles of 120° and 150°. Samples with different hydrophobicity are tested under with $\phi = 0.55$ and $T_s = -7.2$ °C. Experiments show that condensed water droplets are larger with bigger contact angles. No frost appears after 600 s on vertical plate and 140 s on horizontal surface compared with 50 s on normal copper surfaces. Stability is tested, and the surface maintains the superhydrophobicity after ten cycles of frosting and defrosting.

One of the advantages of the superhydrophobic surface is that water droplets are easily removed. Wang et al. [34] show that water droplets fall down easily with contact angle greater than 150°. They propose a technique to construct an aluminum superhydrophobic surface using chemical etching and test the wettability effect on frost formation when spraying on vertical plates. Less surface area is covered by frost on the superhydrophobic aluminum surface. Frost begins to form at some points and increases at these sites, and no new area shows ice formation. The contact angle is measured and shows no significant variation with climatic temperature.

He et al. [37] find that superhydrophobic surfaces retard frost formation at temperatures below the freezing point. Nano-rod arrays of ZnO are fabricated on the test surface. Experiments show that average water SCAs are 170.9°, 166.1°, and 165.8° on the surface with the growth times (t_{ZnO}) of ZnO nano-rod respectively of 1, 2, and 3 h at room temperature. Wettability of the modified surface to condensed micro-droplets is investigated at $T_s = -5$ °C and -10 °C. With $t_{ZnO} = 1$, 2, and 3 h, water SCA = 167.9°, 160.3°, and 154.6° at $T_s = -5$ °C and 163.4°, 158.2° respectively and 155.4° at -10 °C. Solidification time of the condensed droplets is longer on superhydrophobic ZnO nano-rod array surfaces than that on the hydrophobic surface and increases with $t_{ZnO} = 3$, 2, and 1 h. The time is less at -10 °C than that at -5 °C, which is explained by the decrease of nucleation free energy barrier.[8] These results provide the possibility of fabrication of anti-frost materials at low surface temperature conditions.

The durability of a superhydrophobic surface is a challenge to the anti-icing properties of the surface. Farhadi et al. [38] study the anti-icing capability with different superhydrophobic surfaces. Experiments are performed in a wind tunnel by

---

[8]Nucleation free energy barrier is the change in free energy per unit volume and is balanced by the energy gain of creating a new volume and the energy cost due to creation of a new interface.

spraying supercooled water micro-droplets at $-10\ ^\circ$C. Results show that ice adherence increases, and the anti-icing performance of the tested surfaces decreases during icing–deicing cycles. The contact angle decreases and contact angle hysteresis[9] increases on the wet surface caused by water condensation from air.

The theoretical study of the effect of surface wettability on frost formation is limited. Bahadur et al. [39] present a model to predict ice formation on superhydrophobic surfaces resulting from supercooled water droplets. The model is composed of three sub-models: droplet impact dynamics, heat transfer, and heterogeneous ice nucleation. The droplet impact dynamics sub-model predicts the spreading and retraction dynamics of a droplet impacting a superhydrophobic surface. The heat transfer sub-model predicts the transient temperature distribution inside the droplet during spreading and retraction. The heterogeneous ice nucleation sub-model predicts the kinetics of nucleation of ice clusters. The integrated model estimates the droplet radius and retraction force[10] with regard to the retraction time.[11] The model includes the effect of interface temperature on the droplet dynamics and the freezing process. It is validated by experiments with relative humidity of <5%, and the results are in good agreement with the model.

2.2   The Defrost Process

Industrial defrost processes are initiated at intervals to remove the frost layer from working equipment so as to maintain normal operating conditions. Some studies show that surface wettability has an influence on defrosting, and a number of models have been developed to simulate the defrost process. The goal of past investigations was to improve overall defrost efficiency and thus reduce defrost time and energy cost.

Effect of Surface Wettability

Studies of the effect of surface wettability on the defrost process are limited [40–42]. Jhee et al. [41] report the effect of the surface treatment on the performance of the heat exchanger during the frosting and defrosting process. Contact angles on the surfaces tested in their experiments are 12°, 72°, and 124°. During the defrosting stage, the average water drain rates increase 3.7% for hydrophilic heat exchanger and 11.1% for hydrophobic heat exchanger. On the hydrophilic heat exchanger, frost density is higher, and the meltwater absorbed into frost layer is small. The water

---

[9]Contact angle hysteresis is the difference between the advancing and receding contact angles.

[10]Retraction force is the difference in the surface tension forces on the two sides of the droplet.

[11]Retraction time is the time of retraction stage of droplet impact.

drain rate increases accordingly. The water draining rate on the hydrophobic surface has a larger value because frost that is not fully saturated with meltwater drains by gravity. The draining water ratio during the frost melting period is 20.9% for the hydrophilic heat exchanger, 21.5% for the bare one, and 36.2% for the hydrophobic one. The draining rate on the hydrophobic surface is higher as frost drains with meltwater. The average heater temperature on the hydrophilic surface is 4.8 °C lower than the bare one because heat is conducted to melt the frost. The average heater temperature on the hydrophobic surface is higher as only parts of the surface are covered by frost. An interesting observation is that frost accumulation increases by 4.9% for a hydrophilic surface and by 0.4% for a hydrophobic surface. On the hydrophilic surface, the water film increases the coverage area, which promotes frost formation. On the hydrophobic surface, a large blocking ratio[12] increases local air velocity and promotes frost formation. The defrost efficiency is defined as the ratio of the latent heat to the total heat consumption. The defrost efficiency increases by 3.5% for the hydrophilic heat exchanger and by 10.8% for the hydrophobic surface. The enhancement of the efficiency is due to the increment of the water draining rate. It is concluded that the hydrophobic surface has the most significant effect on the defrost process.

Kim and Lee [42] investigate the effect of surface conditions on frosting and defrosting characteristics on an extended surface. Test conditions are set with the $T_a = 2$–6 °C, ambient air velocity 1–2 m/s, $T_s = -10$ to -18 °C, absolute humidity 0.00254–0.00418 kg/kg$_a$, and surfaces with SCA = 2.5°, 75°, and 142° for hydrophilic, bare, and hydrophobic surfaces, respectively. Experiments show that a thin frost film forms on the hydrophilic surface; water droplets and frost crystals appear on the bare surface; and only water droplets form on the hydrophobic surface after 10 min. After 20 min, the three samples have frost on the surface. Frost structures are in different patterns. Pillar-like structures form on the hydrophilic surface, and a hexagonal structure is on hydrophobic surface. Frost thickness is different on the surfaces only during the early stages of frosting. The lowest thickness is on the hydrophilic surface, and the other two are similar. Frost density is highest on the hydrophilic surface, and lowest on hydrophobic surface. Frost surface temperature on the hydrophilic surface is lower than those on the others, but the temperature difference is insignificant. On the hydrophilic surface, the average density is the highest, and thus, the thermal resistance is the lowest. As the frost mass is higher, the heat capacity for defrosting is larger on the hydrophilic surface. On the other hand, the frost density is the lowest on hydrophobic surface, and frost mass is the smallest. The difference in defrosting time is not significant because of the combined influence of frost thermal conductivity and mass of frost layer on the extended surface. The ratio of residual water on the hydrophilic surface is significantly different, which is only 1/6th to 1/10th of that on other fin surfaces. The water on a hydrophilic surface forms a thin film and flows away easily, while water on a hydrophobic

---

[12]Blocking ratio is a parameter indicating the extent of the blockage of the flow passage between the fins of a heat exchanger by frost layer.

surface falls only after forming larger water droplets. Residual water on a hydrophobic surface speeds up frost growth, the frost layer grows faster around the residual water, and frost mass is about 5% greater. Overall the hydrophilic surface exhibits the best thermal performance during testing.

A summary of the effects of the surface wettability on the frost formation and the defrost process from several related investigations is listed in Table 2.1.

Frost Release and Condensate Retention

Surface wettability influences frost release [43–48]. A possibility to release frost from the cold surface is investigated by Wu and Webb [43] with mechanical vibration. They study the effect of surface wettability on frost pattern and melted water and the possibility of frost release on hydrophobic and hydrophilic surfaces.

Experiments are conducted on a vertical plate with $T_a = 24\ °C$, $\phi = 0.55$ and $T_s = -18$ and $-6\ °C$. Frost formed on a hydrophilic surface is uniform in thickness but not uniform on a hydrophobic surface. Condensate droplets are larger on hydrophobic surfaces. Further frost on hydrophobic surfaces cannot be released by surface vibration. During melting, the hydrophilic surface is fully wetted, while the droplets on the hydrophobic surface do not run away, i.e., they tend to stand up on the surface. Surface tension retains some of the condensate and is balanced by gravity and air shear. Condensate retention is influenced by the contact angle of water. It suggests that retained water on the surface should be dried out, otherwise it will refreeze quickly. However, it is very difficult to dry the moisture absorbed into a hydrophilic surface.

Recent literature shows that a fabricated superhydrophobic surface favors the frost release and the defrost process. Jing et al. [46] present an experimental result showing that frost layer detaches from a rigid superhydrophobic surface during the melting. After defrosting, the rigid superhydrophobic surface recovers its superhydrophobic property. Boreyko et al. [47] report a nanostructured superhydrophobic surfaces that promote frost growth in Cassie state.[13] The frost layer is removed by dynamic defrosting which is driven by a low contact angle hysteresis of the meltwater. Chen et al. [48] report on a hierarchical surface that shows influential effect on the frost formation and the defrost stage. The influence rests on suppressing freezing wave propagation during the frost formation and increasing lubrication and mobility of frost during the defrost stage. Frost formation is carried out with $T_a = 22\ °C$ and $\phi = 0.60$. The sample plate is placed horizontally onto a cooling state with present temperature of $-10\ °C$. The study of the condensation evolution shows that droplets freeze beginning from the outer edge corners of the substrate owing to its geometric singularity and low free energy barrier for

---

[13]The Cassie state is the wetting state that air trapped in the grooves between surface features forms a composite (solid–air) hydrophobic surface.

Table 2.1 Wettability effect on the frost–defrost process

	Test sample (L × W × T)	Environmental conditions	Frost/ condensate pattern	Frost layer properties
Shin et al. [27]	100 × 300 × 20 mm Al Advancing DCA: 23°, 55°, 88°	Air: +5 to +20 °C RH: 40–80% Surface: −25 to −5 °C Limitation: Fixed test conditions	Higher DCA: Irregular, dropwise condensate Lower DCA: Uniform, filmwise droplets	Higher DCA: Less dense, larger thickness Lower DCA: Denser, smaller thickness
Liu et al. [30]	150 × 52 × 6 mm Copper Vertical CA: 133°, 63°	Air: +24 °C RH: 62% Surface: −8 °C	Hydrophobic surface: Smaller and fewer water droplets Looser and weaker Plain/Hydro-philic surface: Freezing quicker	The thickness on hydrophobic surface is smaller at the initial stage and is almost the same with that on plain surface after long test
Liu et al. [31]	Copper Horizontal CA: 162°, 72°	Air: 18.4 °C RH: 40% Surface: −10.1 °C		Hydrophobic surface delays frost formation for 55 min and reduces the thickness by 52% at the end
Wang et al. [33]	Copper Horizontal and vertical CA: 120°, 150°	Temperature: −7.2 °C Humidity: 55%	Hydrophobic surface: Condensed droplets become larger when contact angle changes from 120° to 155°. Bare surface: Irregular small droplets	
Wang et al. [34]	20 × 15 × 3 mm Al Vertical CA: 150°	Climatic tem-perature: −6 °C Water tempera-ture: 0 °C (spraying)	Hydrophobic surface: Icing at par-tial area Smooth sur-face: Ice covering the surface	Ice layer thick-ness increases with spraying time

(continued)

Table 2.1 (continued)

	Test sample (L × W × T)	Environmental conditions	Frost/ condensate pattern	Frost layer properties
Kim and Lee [42]	60 × 52.5 × 0.8 mm Al Fin CA: 2.5°, 75°, 142°	Air: +2 to +6 °C Absolute humidity (kg/kg$_a$): 0.00254– 0.00418 Air velocity (m/s): 1.0–2.0 Fin base temperature: −10 to −18 °C	Hydrophobic and bare surfaces: Water droplets Hydrophilic surface: A thin water film	Thickness at early stage: Bare>Hydrophobic > Hydrophilic Frost density: Hydrophilic>Bare>Hydrophobic Defrosting time differs insignificantly
Wu and Webb [43]	40 × 40 × 0.25 mm Al Vertical Advancing/Receding CA: 98°/73° (hydrophobic) 12°/0° (hydrophilic)	Inlet air temperature: 24 °C RH: 55% Surface temperature: −18 to −6 °C	Hydrophobic surface: Nonuniform frost Larger structure Water droplets stay on surface Hydrophilic surface: Uniform frost Fully melted surface during melting	

heterogeneous nucleation. Droplet freezing and the inter-droplet freezing wave vary on hierarchical surface and nano-grassed[14] superhydrophobic surface. On the hierarchical surface with contact angle of ~160° and contact angle hysteresis of 1°, consistent droplets departure causes rare direct freezing of droplets during the condensation stage. The condensate droplets maintain the liquid state until a freezing wave invades from the edge corners at 1410 s. The frozen droplet then sprouts dendritic ice crystal, which is the freezing front, towards the surrounding unfrozen liquid droplets. The newly freezing droplet triggers new ice crystals as well as an inter-droplet freezing wave which eventually propagates over the entire surface. The freezing duration is 395 s. The average propagation velocity of the freezing wave is ~0.9 μm/s. On a nano-grassed superhydrophobic surface with contact angle of ~160° and contact angle hysteresis of ~1–2°, the condensate droplets maintain the liquid

---

[14]The hierarchical surface with nano-grassed micro-truncated cone architecture was fabricated using a combined anisotropic wet-etching and deep reactive ion etching process.

state for 1090 s. The inter-droplet freezing wave propagates at ~1.4 μm/s, which is ~1.5 times that on the hierarchical surface. On the hierarchical superhydrophobic surface, liquid droplets gradually decrease their size by evaporation during the freezing processes. The evaporation of the liquid separates the freezing front from the droplets. All the liquid droplets evaporate without the connection to the freezing front. The success of ice bridging is associated with a length competition between the liquid droplet diameter and frozen ice-to-liquid droplet separation. The ratio of the separation length to the droplet diameter is defined as the bridging parameter. Smaller separation length relative to droplet diameter might cause successful ice bridging. About 85% of liquid droplets on the hierarchical superhydrophobic surface correspond to ice bridging greater than one as opposed to ~67% on the nano-grassed surface. During the defrost stage, bulk frost fracture is observed on a hydrophobic surface. The frost fractures avalanche and create a crack. On the nano-grassed surface, the density of the fracture is low. The frost melts away and leaves a spherical water droplet on the sample surface. In contrast, no visible fracture is found on the hierarchical surface. The frost is detached as a whole from the sample plate during the defrost stage. The properties of frost on hierarchical surfaces show repeatability during many frost and defrost cycles.

Condensate water is retained on the plate surface under defrost conditions. The number of drops and their distribution varies with surface wettability [49–55]. Min and Webb [50] investigate the condensate formation and drainage on the vertical surfaces with different wettability. Substrates are aluminum fin stock, copper fin stock, a coated hydrophilic surface on aluminum, and a coated organic polymer surface on aluminum. Aluminum and copper substrates are treated by acetone cleaning, grinding, and oil contamination separately. The experiments are performed under the air relative humidity of 80–90% and ambient air velocity of 3 m/s. Surface grinding increases the advancing angle and decreases the receding contact angle. The hysteresis is caused by surface heterogeneity, surface roughness and impurities on surface. On the aluminum surface, condensate droplets nucleate at the pits and other imperfections and grow rapidly by vapor condensation and coalescence. After 20 min, an aluminum surface treated by grinding shows no condensate droplets because condensate forms a sheet film. Droplets stand up on a large area of both the aluminum surface and the surface with acetone cleaning. Condensate retention is related to the receding contact angle. Results show that condensate retention increases with the increase of the receding contact angle within 40° because the droplets stand up more. However, as the receding contact angle ranges from 40° to 90°, condensate retention decreases with the increase of the receding contact angle because the surface tension retaining force in vertical direction decreases sharply and size of the droplets decreases. Filmwise condensate forms on the surface with low receding contact angle, while dropwise condensate forms on the surface with high receding contact angle regardless of the advancing contact angle. The height-to-base diameter ratio of the droplets increases with the increasing receding contact angle. When the receding contact angle is very small, condensate fully wets the surface.

EI Sherbini and Jacobi [52] propose a model to predict the amount of condensate retained on plain-fin heat exchangers. The maximum diameter of a retained drop is obtained from the balance between gravitational force and surface tension. The total volume of the drops is obtained by integrating all drop diameters. The prediction of the model agrees well with the measurements by other researchers of the mass of the condensate retained on heat exchangers. Application of the model is restricted to advancing contact angles from 45° to 120°. Microgrooves and surface roughness are found to affect the meltwater retention. Rahman and Jacobi [55] reveal that micro-grooved surfaces drain up to 70% more condensate than a flat surface. During the defrost process, the shape and distribution of the water droplets appear to be random on the plain surface and form parallel streams on the micro-grooved surfaces. Effects of defrost heating rate on the frost surface temperature are investigated. Mass concentration is lower on the micro-grooved surfaces than on the plain surface.

Analytical Models

Defrost can be accomplished using electric resistance heating, hot gas or warm water, and many researchers have developed analytical models to predict thermal performance during the defrost process [56–66]. Defrost on plain (unmodified) surfaces is generally treated, and none of them include the effect of surface wetta-bility on the defrost process.

Sanders [56] presents two models of defrosting process. The first model simulates the hot gas defrosting with constant heat flux, while the second model uses constant defrosting medium temperature to model an electrical defrosting process. Experi-mental results show that defrosting time for the model with constant heat flux is larger than that with constant defrosting medium temperature, so the defrosting efficiency is lower. The model suggests that thick frost layers have enhanced the possibility of occurrence of air gaps. A longer period of defrosting time would be expected while the defrost efficiency might be improved.

Krakow et al. [57, 58] build an idealized model of reversed-cycle hot gas defrosting. The transient defrost cycle in reality is simplified using a quasi-steady-state cycle. When operating at defrost mode, a cold side heat exchanger works as a condenser, and a hot side heat exchanger works as an evaporator. In the defrost model, superheated refrigerant passes through the coil. Experiments show that the surface is not at a uniform condition at a given time. They divide the defrost process into four stages: preheating, melting, vaporizing, and dry heating. The observations show that the frost begins to melt at the refrigerant inlet. The surface conditions are not uniform and might be frosted, slushy, wet or dry at a given time. Part of the resulting water is drained from the surface while the remaining stays as surface water. A glycol-cooled, electrically defrosted coil is used to obtain a uniform condition. In the model, the frost layer is assumed to be a porous medium composed of ice crystals and air. A film of water and entrapped air will form between coil surface and frost layer, and the thickness will increase as defrosting process pro-ceeds. Part of the water will remain due to viscosity, and part will drain due to gravity. During the defrost process, some factors could be indeterminable, such as the initial distribution of the frost, the effect on heat transfer of drained water, and the free-convection heat transfer. This complexity requires an idealized model. The defrost process is modeled by a number of continuous steady state where the time variation is modeled by a large number of successive time intervals. The defrost process is considered as a time-independent steady flow for each time interval. Heat and mass transfer are analyzed for the air-side and refrigerant-side. The refrigerant film conductance is evaluated from correlations. Other heat and mass transfer parameters, including maximum surface water, free-convection air film conductance, air/water film conductance, and surface water vaporization are obtained experimentally.

Sherif and Hertz [59] present a defrost model for a cylindrical coil cooler using electric defrosting method. In the model, total energy released from the electrical heat equals the heat conducted into the melted frost layer and the heat entering into refrigerant vapor. The melt is assumed to continuously drain away from the coil surface. A lumped system analysis for the refrigerant is adopted. Frost–air interface temperature and frost thickness are computed as function of time. The point marking the end of defrosting process is 0 °C corresponding closely to zero-frost thickness. The model has limited application due to the restrictions by the assumptions. It assumes that thermal energy transferred into frost is an arbitrary value. Thermal resistance of refrigerant-coil surface boundary layer and coil thermal resistance are negligible.

Hoffenbecker et al. [62] develop a transient model to predict defrost time and efficiency. A unique contribution of the model is to estimate the parasitic energy associated with thermal convection, moisture re-evaporation, and extraction of the stored energy in the coil. Heat and mass transfer mechanisms during the defrost process include condensation of high-pressure high temperature gaseous refrigerant inside tubes, heat conduction through coil tubes and fins, sensible heating of accumulated frost, latent melting of accumulated frost, and re-evaporation of moisture from coil surface to surrounding. A finite difference approach is implemented, and an implicit time integration scheme is applied. One difficulty is the effect of phase change on the continuity of the computational domain. Instead of solving a moving boundary domain, the model approximates the effects of frost melting by assuming mass and volume at each node are constant. When ice melts, water drains, and the volume formally occupied by water is replaced by air. The model over-estimates the mass-specific heat product, but the simulation is supported by comparisons with experimental results and the results of other studies. Experimental information is obtained from a fruit product storage freezer. The defrost time agrees with the predicted value. Only 43.7% of total energy input is used for melting the frost. Results show that the initial rate of energy input is high and decreases with time. The model predicts that the mass of moisture re-evaporated back to the refrigerated space increases with decreasing hot gas temperature due to prolonged defrost dwell time. Parasitic energy impacts could be minimized by limiting defrost dwell time. The model suggests that an optimum hot gas temperature is a function of both accumulated mass and density of frost.

Dopazo et al. [63] model the defrost process with six stages: preheating, tube frost melting start, fin frost melting start, air presence, tube–fin water film, and dry-heating. Assumptions include the following: refrigerant liquid and vapor phases are in thermodynamic equilibrium, constant thermal properties of the refrigerant and the tube wall, and insignificant heat transfer in axial direction. Heat and mass transfer and energy equation were analyzed. Convective heat transfer coefficients, condensation heat transfer coefficient, free convection coefficient, and convective mass transfer coefficient are obtained from current correlations. A finite difference method is used to find the defrost time and energy distribution during the defrost process as well as temperature distribution on tube, fin, and frost. The simulation results agree with the experimental results regarding to the defrost time and the energy used to melt the frost. The influence of mass flow rate and inlet temperature on defrost time, and energy delivered by the refrigerant is also studied in the model. Experiments show that defrost time and energy supplied by the refrigerant increases with the decreasing mass flow rate.

An experimental study of the reverse cycle defrosting performance of a multi-circuit outdoor coil[15] on a 6.5 kW air source heat pump is reported by Minglu et al. [64]. The outdoor coil includes four refrigerant circuits on a single tube. An environmental chamber with a separate air conditioning system is provided to control the temperature and humidity. The temperature of outdoor ambient is 0.5 °C, and $\phi = 0.90$. After 2 h of frosting, reverse cycle defrosting starts. Results show that no solid frost left after 7.5 min into defrosting. Frost melting on the upper level coil is quicker than that on the lower surface of outdoor coil. Surface temperature on the upper level circuit is larger than that on the lower level. A noticeable phenomenon is that melted frost flows downwards by gravity from high level along coil surface of lower circuit. The moving frost column causes a negative effect on the defrosting efficiency. As melted frost temperature is significantly lower than coil surface temperature, part of the heat is taken away by the moving frost column, which results in the increase of water temperature in the collecting pan. The moving frost column also brings greater thermal load to the lower level circuits. Total heat input increases with the delayed termination time, which is determined by the surface temperature of the lowest level circuit. Thus defrosting efficiency is reduced due to the effect of the melted frost flowing. Qu et al. [65] also propose an analytical model. The defrosting process is divided into three stages: frost melting without water flow, frost melting with water flow and water layer vaporization. The melted frost is held to the surface at first due to surface tension until the mass of the melted frost held reaches the maximum point and then flows downwards due to gravity. The outdoor coil surface is divided into four control volumes, and lumped parameter modeling is applied. The mass flow rate of refrigerant is assumed to be evenly distributed into the four refrigerant circuits. The water layer is considered to be in the laminar regime due to the very small velocity observed from the experiments. The contact area between the melted frost and the frost layer increases with flowing

---

[15]The term "coil" refers to low temperature, fin-tube heat exchanger surfaces.

water. The flow resistance increases downwards, and the velocity of water layer decreases from the top to the bottom circuits. Prediction of the defrost duration and temperature variation of the collected melted frost agrees well with the experimental results. Mass flow of the melted frost increases while the frost melting rate decreases from the top circuit to the bottom circuit.

Mohs [66] proposes a defrosting model including vapor diffusion. The defrost process is divided into the vapor diffusion sub-model, permeation sub-model, and dry out sub-model. The initial frost is composed of ice crystals and air pockets. The porosity could be constant or variable. The water vapor pressure within air pockets is assumed to be at the saturation pressure of the local frost temperature. During first stage, water molecules diffuse from the ice surface, through the frost layer, and escape into the surrounding air due to the increasing vapor pressure. Thus, latent heat and sensible heat is lost to the surrounding air. In the second stage, permeation, the composite of a layer of meltwater and ice is formed. Meltwater is absorbed into frost layer by capillary forces. Within the permeation layer, the water content varies from completely water near the surface to no water further into the frost layer. Two fronts may be expected. The permeation layer is moving from the surface, while the air/frost front is moving towards the surface. After the second stage, a small portion of liquid will adhere to the surface through surface tension. Defrost efficiency during the evaporation process is very low because most of the supplied heat are lost into the ambient air. The amount of retained liquid mass on the surface will be a function of inclination angle and surface wettability. The model includes a vaporization phase at the beginning of the defrost process and can be used to estimate defrost rate and efficiency.

Ice Adhesion

Adhesive forces attach the frost column to the solid surface, affects by temperature and surface wettability. However bulk movement of the frost column is possible during the defrost process and is determined by the balance of forces acting on the column.

Raraty and Tabor [67] investigated the adhesion of ice to various solids. The interfacial strength of ice to metals is larger than the strength of ice. Their experiments show that the primary factor causing failure is the rate of strain at the interface instead of the amount of strain. When tensile stress is high, the failure is brittle, and the breaking stress is independent of temperature. If the tensile stress is below a critical limit, the failure is ductile, and the breaking stress increases linearly as the temperature is reduced below 0 °C. Ductile failure appears to be determined by the onset of a critical creep rate. Ice adhesion is reduced by a large factor with surface contaminants on metals. This is caused by the reduction of contact area over which strong metal/ice adhesion occurs. The interfacial strength of ice on polymeric materials is lower than the strength of ice, thus failure occurs at the interface. Shear strength of the solid–ice interface is studied with polymers sliding on ice.

The study shows that ice layers may be removed more easily if brittle fracture can be achieved. Constraints of ice inhibit brittle fracture. Hydrophobic materials show a very low adhesion.

Jellinek [68] studies the adhesive strength of ice by shear experiments. Measurements show that snow-ice on stainless steel yields adhesive breaks to a temperature of about −13 °C, where a sharp transition to cohesive breaks take place. Ice on polystyrene yields pure adhesive breaks on shear. The adhesive strength is found to be a function of temperature. A marked difference is found in the behavior of the ice disks in tension and in shear. The experimental results are explained by assuming a liquid-like layer between ice and the surface. The thickness and consistency of the layer are functions of temperature and the nature of the surface.

Ryzhkin and Petrenko [69] present an electrostatic model of ice adhesion. When distances are greater than one intermolecular distance, the model gives an order of magnitude for the adhesive energy, which is significantly greater than both chemical bonding energy and van der Waals forces.

Makkonen [70] presents concepts and models to estimate ice adhesion. The material deformations affect adhesion strength, which is explained by an indirect effect arising from the brittle-ductile nature of ice. The ice interface behaves as a ductile material with little tendency to fracture even at high strain rates when the substrate material is either highly elastic or the ice is close to its melting temperature. On a rigid substrate or cold surface, however, mechanical removal of ice occurs in a brittle manner. Interface morphology and crystal structure of ice have effects on adhesion strength. When the interface includes micro-pores filled with water, the true interface area is large. In such case, a pure adhesion-failure might not be possible, and fracture occurs partly due to cohesion-failure within the ice. The effects of temperature and substrate materials on ice adhesion are discussed. A liquid film at the solid–ice interface is expected to reduce adhesive strength. The thicker the liquid film, the lower the adhesion is expected. Qualitatively, the adhesion strength of ice increases with decreasing temperature down to the temperature at which the liquid film disappears.

Chen et al. [71] present a study on ice adhesion on surfaces of different wettability. Experimental results show that a superhydrophobic surface cannot reduce the ice adhesion, and ice adhesion strength on the superhydrophilic surface is almost the same as that on the superhydrophobic surface. This phenomenon is explained by the mechanical interlocking between ice and the surface texture. Meuler et al. [72] present a relationship between water wettability and ice adhesion. Twenty-two surfaces with different advancing/receding contact angles are tested. The strengths of ice adhesion are measured on bare steel discs. A liquid-like interface between ice and a substrate could facilitate lateral sliding prior to detachment of the ice column. The interactions of the substrate with liquid water are scaled with respect to liquid drop roll-off angle, equilibrium work of adhesion and practical work of adhesion. A strong correlation is found between the ice adhesion and the practical work of adhesion.

The instability of an ultra-thin water film on a hydrophilic surface is investigated by Majumdar and Mezic [73], who propose a new theory of droplet formation during condensation. The theory uses hydration, electrostatic force and van der Waals force between a hydrophilic solid surface and a water film. Hydration force dominates for films thinner than ~3 nm. At 300 K, the equilibrium film thickness is found to be almost constant at 0.5 nm for a wide range of relative humidity and increases sharply when the relative humidity approaches to unity. The theory shows that the competition of strain energy with hydration, van der Waals force, and liquid–vapor surface tension induces the instability for films thicker than a critical value.

2.3 Summary

A wide range of studies have been conducted on frost formation and the defrost process. The effect of surface wettability has attracted a lot of interests for decades. The study on the effect of surface wettability has been focused mainly on frost formation. For the defrost process, the study is focused on the relationship between retention droplets and water contact angles of substrates. There are a few papers in the literature that present the influence of contact angle on defrost time and efficiency. The study of frost removal attracts the interests of researchers with the findings of surface characteristics. The superhydrophobic surface might be able to facilitate the defrost process by reducing the evaporation time of water retention, and defrost efficiency could improve accordingly.

Chapter 3
A Model of the Defrost Process

3.1 General Description of the Melting Model

The defrost process includes surface heating, melting of the frost layer, and vaporization of the retained water. Melting behavior has been widely studied in cold region science and water resources research, and the complexity lies in that the permeation of the meltwater is not easily defined. The permeation rate depends on frost structure, surface tension, capillary pressure, and water saturation.[1]

When the frost layer is heated at the solid surface, meltwater either is absorbed into the frost layer or drains. To fully investigate this process, we develop a melting model in this chapter to analyze the physical behavior of the meltwater. The model has three stages: absorption, accumulation, and draining. Melting begins with absorption of the meltwater. The absorption process depends on surface tension, fluid viscosity, grain size, porosity, and capillary pressure. When the frost layer becomes saturated with meltwater, accumulation occurs wherein a thin water film accumulates between the solid surface and the permeation layer. The water film is balanced by gravity, surface tension, and shear force. With the growth of the water film, the meltwater drains by gravity.

The composition of the frost layer varies with the stages of the melting process. When the permeation rate of meltwater is greater than the melting rate, all the meltwater is absorbed into the frost layer during the absorption stage. The mass composition consists of a permeation and frost layer. The meltwater accumulates between the surface and the permeation layer when the frost layer is saturated. The mass composition includes a thin water film and the permeation layer. A summary of the mass composition is listed in Table 3.1.

The melting process is closely related to surface wettability. In the absorption stage, the permeation rate is determined by the porous structure of the frost which is a

---

[1] Water saturation is the volume fraction of water in the pore volume.

© The Author(s), under exclusive license to Springer Nature Switzerland AG 2019
Y. Liu, F. A. Kulacki, *The Effect of Surface Wettability on the Defrost Process*,
SpringerBriefs in Applied Sciences and Technology,
https://doi.org/10.1007/978-3-030-02616-5_3

Table 3.1 Physical processes and mass composition during melting

Melting process	Water film without draining	Water film with draining	Permeation layer	Frost layer
I Absorption			▓	▓
II Accumulation	▓		▓	
III Drainage		▓	▓	

Fig. 3.1 A general illustration of the initial frost column

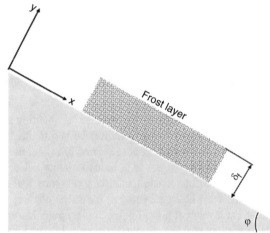

function of surface wettability in the frost formation period. The effect of surface wettability in the accumulation stage lies in that the maximum thickness of the thin water film depends on the surface tension acting on the perimeter of the frost column, and the surface tension is related to surface wettability. When the meltwater drains by gravity, the average volumetric flow depends on the velocity boundary conditions at the solid–water interface and the water–permeation interface. A slip boundary condition for velocity applies at the solid–water interface for hydrophobic surface.

The melting process is formulated in terms of the key variables, or physical quantities in each stage. In the absorption stage, the permeation layer is a mixture of ice crystals, water and air. The rate of meltwater permeation is expressed in the mass continuity of the water saturation. In the accumulation stage, the maximum water thickness is expressed by a static force analysis on the water film. The growth of the water film depends on the energy transferred into the permeation layer. In the drainage stage, the draining velocity is expressed by the momentum equation of the draining water.

The initial mass of the frost layer comprises ice crystals and air. At the beginning of the melting process, the temperature across the frost layer is assumed to be uniform. Heat transfer is assumed to be one-dimensional and normal to the test surface. A general diagram of the frost layer structure is shown in Fig. 3.1. The initial frost thickness is δ_f. The direction along the surface is x, and the direction normal to the surface is y. A special condition considered here is where the slope angle ϕ is 90°, and the surface is placed vertically.

3.2 Meltwater Absorption

Absorption of meltwater occurs at the beginning of the melting process. The assumptions for the absorption process include the following: (1) the permeation rate is greater than the melting rate of the frost layer, and all meltwater is absorbed into the frost layer; (2) the frost layer is treated as rigid porous medium; (3) temperature in the permeation layer remains at the melting temperature; (4) capillary pressure varies with the water saturation; and (5) water saturation equals unity at the solid–permeation interface and zero at the permeation–frost interface.

The bulk frost column thus comprises the permeation layer and the frost layer. A control volume (CV) analysis is applied to formulate the mass and heat transfer equations in the permeation and frost layers. The CV is selected in the permeation layer for the mass balance analysis (Fig. 3.2).

The mass of water in the CV is written as a function of the water saturation S,

$$m_w = \varepsilon \rho_w S A_{ts} dy \tag{3.1}$$

The mass balance of meltwater in the permeation layer is

$$\varepsilon \rho_w \frac{\partial S}{\partial t} = -\frac{\partial m''_{w,p}}{\partial y} - \varepsilon \rho_w v_f \frac{\partial S}{\partial y} \tag{3.2}$$

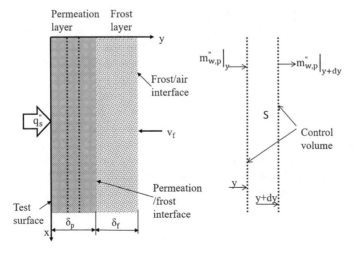

Fig. 3.2 Control volume in the permeation layer during absorption of meltwater

where the LHS is the rate of change of water mass in the CV, the first term on the RHS is the net mass transfer into the CV by water permeation, and the second term on the RHS is the mass transport due to the motion of the bulk mass.

The meltwater velocity depends on the heat flux applied at the test surface, the density of the frost layer and the latent heat of fusion, L_f,

$$v_f = -\frac{q''_s}{L_f \rho_f} \tag{3.3}$$

where

$$\rho_f \approx (1 - \varepsilon)\rho_i \tag{3.4}$$

The minus sign indicates that the direction of the movement is towards the test surface. The permeation layer is a mixture of air, water and ice crystals. At the solid surface, the permeation layer contains meltwater and ice crystals and is maintained at the melting temperature, thus only the latent heat is considered.

The meltwater permeation flux, $m''_{w,p} = \rho_w v_w$, where v_w is the volume flux of water, and Eq. (3.2) becomes,

$$\varepsilon \frac{\partial S}{\partial t} = -\frac{\partial v_w}{\partial y} - \varepsilon v_f \frac{\partial S}{\partial y} \tag{3.5}$$

The volume flux of the meltwater is given by Darcy's law,

$$v_w = \frac{K_w}{\mu_w} \frac{\partial P_{ca}}{\partial y} \tag{3.6}$$

In a porous medium such as frost, the meltwater volume flux depends on capillary pressure, fluid viscosity, surface tension, and structure of the medium [74–89]. The water flow is generally driven by the gradient of the capillary pressure, which is a function of water saturation. The structural description of the porous medium includes grain size, tortuosity, and porosity. The rate of liquid penetration in a porous body may be modeled as a number of cylindrical capillary tubes with varying radii and is proportional to the square root of time and to the square root of the ratio of the surface tension to gravity. Aoki et al. [74] present a model describing the melting process of a snow layer when heat is supplied at the bottom and show that the melt time and the melting efficiency are affected by water permeation. The calculated melting efficiency is higher when neglecting water permeation. An empirical correlation is applied to represent the water permeation rate, which relates to the capillary suction pressure and the porous structure. Their empirical correlation for the dimensionless capillary pressure is

$$P^+(S) = \frac{P}{\gamma} \frac{\varepsilon}{(1-\varepsilon)} \frac{D}{6} = 1.6e^{-0.46S} - e^{20(S-1)} \tag{3.7}$$

Colbeck [75–78] presents the capillary effects on water percolation and investigates the relation of capillary pressure to the water saturation. In Colbeck's model, gravity is the driving force for water flow. Capillarity accounts for less than 10% of the total forces when the water volume flux is 10^{-8} m/s. The relation between capillary pressure and water saturation depends on a number of parameters including contact angle, interfacial tension, particle shape, and pore-size distribution. A typical expression [81] relating effective permeability and the capillary pressure is by a power law, where the permeability, K_w, and capillary pressure, P_{ca}, are assumed to be functions of the water saturation, S, and are given as

$$K_w = K_w(S) = KS^p, \quad P_{ca} = P_{ca}(S) = CS^{-q} \tag{3.8}$$

where K, C, p, and q are constants, K is the intrinsic permeability (a function of the geometry of the porous medium), and C is the entry capillary pressure. The exponent q is related to the pore-size distribution. The volume flux of meltwater thus becomes,

$$v_w = -\frac{KCq}{\mu_w} S^{p-q-1} \frac{\partial S}{\partial y} \tag{3.9}$$

Substituting Eq. (3.9) into Eq. (3.5), the meltwater mass balance becomes

$$\varepsilon \frac{\partial S}{\partial t} = \frac{KCq}{\mu_w} \left[(p - q - 1)S^{p-q-2} \left(\frac{\partial S}{\partial y} \right)^2 + S^{p-q-1} \frac{\partial^2 S}{\partial y^2} \right] - \varepsilon v_f \frac{\partial S}{\partial y} \tag{3.10}$$

Within the permeation layer, water saturation decreases from the solid–water interface to the permeation front. Thus water content at the solid–permeation interface and at the permeation–frost interface is

$$y = 0, \quad S = 1 \tag{3.11a}$$

$$y = \delta_p, \quad S = 0 \tag{3.11b}$$

The initial condition is

$$t = 0, \quad S = 0 \tag{3.11c}$$

The temperature of the permeation layer is assumed to be constant at the melting temperature. A CV is selected in the frost layer for the energy analysis (Fig. 3.3). The one-dimensional energy balance for the frost layer is

$$(\rho c_p)_f \frac{\partial T_f}{\partial t} = \frac{\partial}{\partial y} \left(k_f \frac{\partial T_f}{\partial y} \right) - (\rho c_p)_f v_f \frac{\partial T_f}{\partial y} \tag{3.12}$$

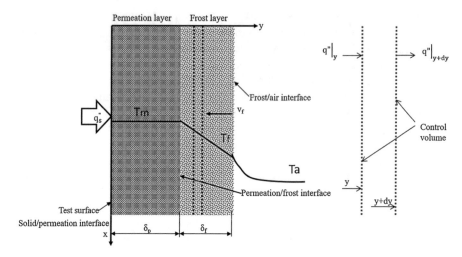

Fig. 3.3 Control volume in the frost layer during the absorption

where the LHS represents the time rate of change of energy in the CV, the first term on the RHS represents heat transfer by conduction, and the second term on the RHS represents the energy transport due to bulk motion. The boundary conditions at the solid–permeation interface, permeation–frost interface, and frost–air interface are as follows:

$$y = 0, \qquad q'' = q''_s, \tag{3.13a}$$

$$y = \delta_p, \qquad T_f = T_m, \tag{3.13b}$$

$$y = \delta_{p+f}, \quad -k_f \frac{\partial T_f}{\partial y} = h_a(T_f - T_a), \tag{3.13c}$$

$$t = 0, \qquad T_f = T_m. \tag{3.13d}$$

3.3   Meltwater Accumulation

When the frost layer is saturated, a water film accumulates between the solid surface and the permeation layer. The frost column then comprises a water film and permeation layer (Fig. 3.4).

The growth of the water film is formulated from the energy balance analysis at the water–permeation interface.

$$-k_w \frac{\partial T_w}{\partial y} \bigg|_{\delta_w} = (1 - \varepsilon)\rho_i L_f v_y, \tag{3.14}$$

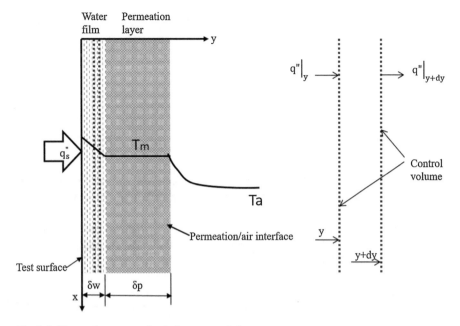

Fig. 3.4 Heat and mass transfer during accumulation

where v_y is the velocity of the interface. The interface velocity of the permeation layer is

$$v_y = \frac{d\delta_w}{dt}.$$

(3.15)

Substituting Eq. (3.15) into Eq. (3.14), the energy equation at the water–permeation interface becomes

$$-k_w \frac{\partial T_w}{\partial y}\bigg|_{\delta_w} = (1 - \varepsilon)\rho_i L_f \frac{d\delta_w}{dt}$$

(3.16)

The energy balance in the water film is:

$$(\rho c_p)_w \frac{\partial T_w}{\partial t} = \frac{\partial}{\partial y}\left(k_w \frac{\partial T_w}{\partial y}\right)$$

(3.17)

The boundary and initial conditions at the solid–water interface and the water–permeation interface are:

$$y = 0, \qquad q'' = q''_s \tag{3.18a}$$

$$y = \delta_w, \qquad T_w = T_p = T_m, \tag{3.18b}$$

$$t = 0, \qquad T_w = T_m. \tag{3.18c}$$

3.4   Meltwater Draining

Meltwater drains away when the frost layer is saturated and when the gravity outweighs the adhesive force at the solid–water interface. A draining model is shown in Fig. 3.5.

Assuming that the meltwater drains between the test plate and frost column, the draining velocity is determined by the boundary conditions at the interfaces.

The momentum balance in the x-direction along the surface is

$$\rho_w \left(\frac{\partial u_w}{\partial t} + u_w \frac{\partial u_w}{\partial x} + v_w \frac{\partial u_w}{\partial y} \right) = \rho_w g_x + \mu_w \left(\frac{\partial^2 u_w}{\partial x^2} + \frac{\partial^2 u_w}{\partial y^2} \right) - \frac{\partial P}{\partial x} \tag{3.19}$$

The momentum balance in y-direction is

$$\rho_w \left(\frac{\partial v_w}{\partial t} + u_w \frac{\partial v_w}{\partial x} + v_w \frac{\partial v_w}{\partial y} \right) = \rho_w g_y + \mu_w \left(\frac{\partial^2 v_w}{\partial x^2} + \frac{\partial^2 v_w}{\partial y^2} \right) - \frac{\partial P}{\partial y} \tag{3.20}$$

$$g_x = g \sin \varphi, \quad g_y = -g \cos \varphi \tag{3.21}$$

Fig. 3.5 Drainage model

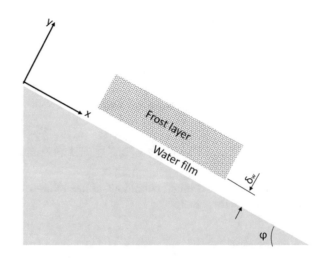

The meltwater is assumed to drain along the x-direction and is only a function of y. Thus Eqs. (3.19) and (3.20) are simplified as follows:

$$\frac{\partial u_{w}}{\partial t} + v_{w}\frac{\partial u_{w}}{\partial y} = \mu_{w}\frac{\partial^{2} u_{w}}{\partial y^{2}} + g\sin\varphi. \tag{3.22}$$

$$\frac{\partial P}{\partial y} = -\rho_{w}g\cos\varphi. \tag{3.23}$$

When $\varphi = 90°$, the frost surface is vertical, and the draining model is shown in Fig. 3.6. The energy equation is similar to that in the accumulation stage. The draining velocity varies in the y-direction and is perpendicular to the solid plate.

At the water–permeation interface, a slip boundary condition is applied owing to the effects of viscous shear [90–93]. Beavers and Joseph [90] present a slip velocity proportional to the shear rate at the permeable boundary for free fluid over a porous medium,

$$y = \delta_{w}, \qquad \frac{du_{w}}{dy} = \frac{\beta}{\sqrt{K}}(u_{B} - \omega) \tag{3.24}$$

where K is the permeability of the permeation layer, β is a dimensionless parameter depending on the material which characterizes the structure of permeable material within the boundary region, ω is the filtering velocity of water in the permeation layer with a unit of m/s, and u_{B} is the velocity at the water–permeation interface.

For viscous liquids at low Reynolds number, flow in permeable material is governed by Darcy's law. When gravity dominates the flow, ω is:

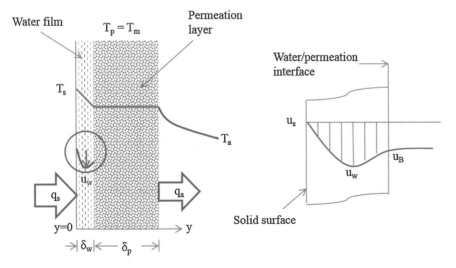

Fig. 3.6 Heat and mass transfer during drainage for a vertical frost surface

Fig. 3.7 Definition of the
slip length for a simple shear
flow [98]

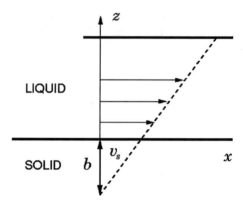

$$\omega = \frac{K}{\mu_w}\rho_w g \qquad (3.25)$$

At the solid–water interface, boundary conditions depend on the surface wettability. For a hydrophilic surface, a no-slip boundary condition is assumed at the
solid–water interface,

$$y = 0, \qquad\qquad u_w = 0 \qquad (3.26)$$

For a thin liquid film, velocity slip is present for flow over hydrophobic surfaces
[94–100]. A slip velocity condition is therefore applied to the hydrophobic surface.
The slip velocity, u_s, on the wall can be expressed [98] as follows:

$$u_s = b\frac{\partial u_w}{\partial y} \qquad (3.27)$$

where b is the slip length and can be estimated as

$$b = \delta\left(\frac{\mu_b}{\mu_s} - 1\right) \qquad (3.28)$$

where δ is the thickness of the thin liquid layer at the boundary, μ_b is the bulk
viscosity, and μ_s is the average viscosity of the near-to-wall layer. The slip length is
defined as an extrapolated distance relative to the wall where the tangential velocity
vanishes as shown in Fig. 3.7.

The average draining velocity across the water film thickness is

$$u_{avg} = \frac{\int_0^{\delta_w} u_w dy}{\delta_w} \qquad (3.29)$$

And the average volumetric flow rate across is

$$\Omega = u_{\mathrm{avg}} W \delta_{\mathrm{w}} \tag{3.30}$$

where W is the width of the surface.

3.5 Defrost Time and Efficiency

Defrost time and efficiency are used to evaluate the system effectiveness. Both depend on system design, ambient temperature, and heating methods. Defrost time includes preheating time, melting time of the frost column, draining time of the meltwater, and evaporation time of the retained water. Defrost time for a lumped frost layer can be defined as follows:

$$t_{\mathrm{d}f} = \frac{(1-\varepsilon)\rho_{\mathrm{i}}\delta_{\mathrm{f}}L_{\mathrm{f}}}{q''_{\mathrm{s}} - h_{\mathrm{a}}(T_{\mathrm{m}} - T_{\mathrm{a}})} + \frac{(1-\varepsilon)\rho_{\mathrm{i}}\delta_{\mathrm{f}}H}{\rho_{\mathrm{w}}u_{\mathrm{avg}}\delta_{\mathrm{w}}} \tag{3.31}$$

where H is the height of the solid surface. Defrost time in the definition includes the melting time of the frost column and the draining time of the meltwater. Defrost efficiency is defined as the ratio of the energy required to melt the frost to the total energy applied at the surface,

$$\eta_{\mathrm{d}f} = \frac{(1-\varepsilon)\rho_{\mathrm{i}}\delta_{\mathrm{f}}\left[L_{\mathrm{f}} + c_{\mathrm{p,i}}\Delta T_{\mathrm{f}}\right]}{q''_{\mathrm{s}}t_{\mathrm{d}f}} \tag{3.32}$$

and is a function of the frost mass, the initial temperature of the frost layer, the heat flux applied at the test surface, and the defrost time. Defrost efficiency also varies with different melting stages.

3.6 Slumping Criteria

Slumping is a process of frost removal that is different from the traditional melting process. Slumping occurs when the frost layer either peels or breaks away in large chunks from the surface and moves in bulk along or away from the surface. Unbalanced forces might also cause the frost layer to break up into pieces that fall from the surface during the defrost process. When subject to axial compression, the frost column might buckle and fail. At low temperature, thermal contraction or expansion might cause failure of the solid–frost interface, which is due to the different thermal expansion coefficients of the solid and the ice crystals. The frost column and the solid surface may also cool at different rates when their heat capacities and thermal conductivities differ. Additional stress at the interface results from this. The significance of the added stress depends on the cooling rate. When a

Table 3.2 Failure modes of frost layer

Melting periods	Possible Failure modes			
	Static force balance	Instability of water film	Buckling due to axial compression	Thermal expansion/contraction
Pre-melting		Water thickness is less than 100 nm. Frost column might break up into pieces		
Absorption	Frost column falls off from the solid surface			
Accumulation	Frost column falls off from the solid surface			
Drainage	Frost column falls off from the solid surface			
On site coils with complex geometry			Frost column buckles when subjected to compression	
Sudden cooling				Failure at the interface due to different thermal expansion coefficients of solid and ice

thin liquid film exists at the solid surface, the interface might be ductile, and it would adjust to the thermal expansion without fracturing. In the best situation, the frost layer falls as a whole rigid piece. Time for water draining and evaporation would be eliminated, and thus defrost efficiency would improve. The possible criteria of frost failure are attributed to several failure categories are shown in Table 3.2.

A static force balance is applied to formulate the slumping criterion for each of the melting stages. Frost slumping is likely to occur when the adhesive force and surface tension get weak. Water accumulation between the test surface and the frost column is a favorable condition that weakens the adhesive force. When gravity outweighs adhesive force or surface tension, the frost layer peels off from the solid surface during the melting process. The different failure mechanisms are illustrated in Fig. 3.8 for hydrophilic and hydrophobic surfaces. At the beginning of melting, the force balance is determined by the competition between the gravity and adhesion force. When the frost layer is saturated with the meltwater, a thin water film develops

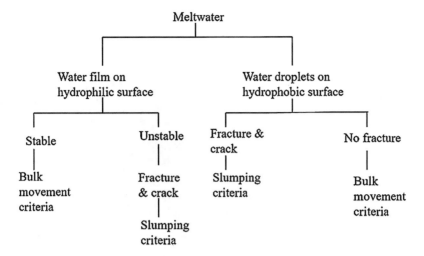

Fig. 3.8 The scheme and mechanisms of frost slumping

and is subjected to gravity and shear at its interfaces. On hydrophilic surfaces, meltwater is assumed in the form of thin water film. When the water film becomes unstable, it breaks up and causes fracture in the frost column. On hydrophobic surfaces, meltwater is assumed in the form of water droplets. During melting, the adhesive force at the solid–frost interface and the cohesive force inside the frost layer might become unbalanced, which causes fracture and cracking of the frost column.

Force Balance in the Absorption Process

During meltwater absorption, when the permeation rate is greater than the melting rate, all the meltwater is absorbed into the frost layer. A permeation layer is formed between the solid surface and the frost layer as shown in Fig. 3.9. The bulk column is subjected to gravity and adhesion force at the solid–permeation interface. When gravity exceeds adhesion in shear, the frost column might pull away from the solid surface. The body force on the frost column varies with the porosity and thickness of the frost layer. The adhesion force in shear is determined by surface wettability and is also dependent on the surface temperature.

The force balance on the bulk column is

$$m_{f0}g - \tau_i A_s = 0 \tag{3.33}$$

where m_{f0} is the mass of the bulk column,

$$m_{f0} \approx A_s(1 - \varepsilon)\delta_{f0}\rho_i \tag{3.34}$$

Fig. 3.9 Force balance
during absorption

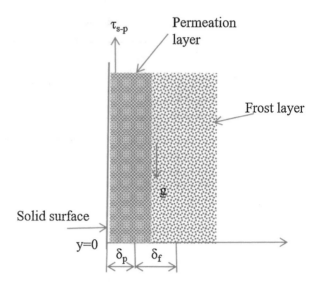

Adhesion under shear is much smaller than that under tension because a liquid-like layer is assumed to exist between the ice crystals and the solid surface. The shear strength of ice adhesion is related to the advancing, or receding, contact angle. The correlation between ice adhesion, τ_i, and the contact angle [72] is

$$\tau_i = (340 \pm 40)(1 + \cos\theta_{rec}) \quad [\text{kPa}], \quad \text{Residual} = 0.92, \qquad (3.35)$$

where the ice adhesion is measured at $-10\,°\text{C}$.

The adhesive force also decreases with the increase of the temperature. As the surface temperature increases, gravity then dominates and can lead to frost slumping. The strength of solid–liquid interaction is greater on hydrophilic surfaces than on hydrophobic surfaces. Thus hydrophilic surfaces could be more favorable to slumping.

The Force Balance During Meltwater Accumulation

During meltwater accumulation, when the frost layer is either saturated with melt-water or under the condition that the melting rate is greater than the permeation rate, meltwater can accumulate at the solid surface. A thin water film then stands between the solid surface and the frost layer (Fig. 3.10). The maximum thickness of the water film is determined by a force balance. The water film is subject to gravity and shear forces at the solid–water and water–frost interfaces. The frost column is subject to the body force and the shear force at the water–frost interface. The bulk mass is attached to the solid surface by surface tension. The shear force at the solid surface is

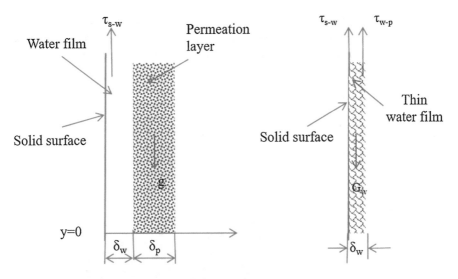

Fig. 3.10 Force balance during accumulation

associated with the surface wettability and surface temperature. The force balance for water film is

$$G_w - F_{s-w} - F_{w-p} = 0 \tag{3.36}$$

The force balance for frost column is

$$G_f + F_{w-p} = 0 \tag{3.37}$$

On hydrophilic surfaces, force balance on the bulk mass becomes,

$$m_{f0}g = 2\gamma(L + W)\cos\theta \tag{3.38}$$

where γ is the surface tension of water, L is the length, W is the width of the test surface, and θ is the water contact angle at the solid surface. On hydrophobic surfaces, surface tension is related to the advancing contact angle θ_{adv} and the receding contact angle θ_{rec}[2] as shown in Fig. 3.11.

The force balance on the hydrophobic surface is

$$m_{f0}g = 2\gamma(L + W)(\cos\theta_{rec} - \cos\theta_{adv}) \tag{3.39}$$

---

[2]The contact angles formed by expanding and contracting the liquids are referred to as the advancing contact angle and the receding contact angle respectively.

The difference between the advancing contact angle and the receding contact angle is called contact angle hysteresis. Hysteresis is caused by surface roughness, heterogeneity and impurities on surface. Surface tension is a function of contact angle hysteresis.

Fig. 3.11 Surface tension on hydrophobic surfaces

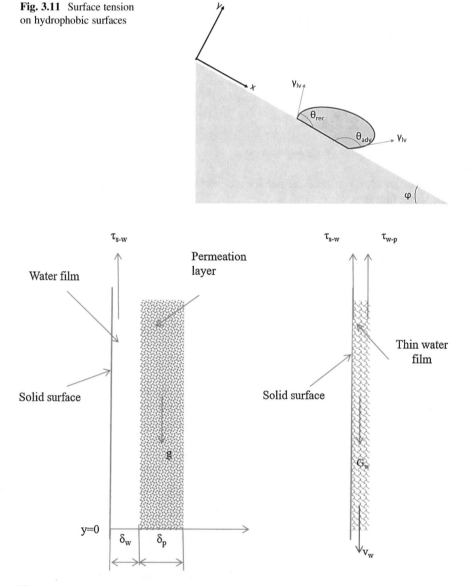

Fig. 3.12 Force balance during drainage

The Force Balance During Meltwater Draining

Meltwater draining occurs by the body force in the presence of the shear force at the solid surface as shown in Fig. 3.12. The body force decreases with draining, and the shear force also decreases with the rising surface temperature. Slumping would be possible when the decrease of the shear stress is much larger than the decrease of the body force. The bulk frost mass then falls off with the draining water.

Chapter 4
Solution Methods

4.1 Absorption

During absorption of meltwater, permeation rate depends on saturation distribution in the permeation layer and is governed by Eq. (3.10). Front fixing methods are applied to track the phase front as the phase boundary moves [101]. Letting $y^+ = \frac{y}{\delta_p(t)}$, the dimensionless form of the equation for meltwater saturation becomes

$$\frac{\partial S(t,y)}{\partial y} = \frac{\partial S}{\partial y^+} \frac{\partial y^+}{\partial y} = \frac{1}{\delta_p(t)} \frac{\partial S(t,y^+)}{\partial y^+} \tag{4.1}$$

$$\frac{\partial^2 S(t,y)}{\partial y^2} = \frac{\partial S}{\partial y}\left(\frac{\partial S}{\partial y}\right) = \frac{1}{\delta_p(t)^2} \frac{\partial^2 S(t,y^+)}{\partial y^{+2}} \tag{4.2}$$

$$\frac{\partial S(t,y)}{\partial t} = \frac{\partial S(t,y^+)}{\partial t} - \frac{y^+}{\delta_p(t)} \frac{d\delta_p(t)}{dt} \frac{\partial S(t,y^+)}{\partial y^+}. \tag{4.3}$$

The dimensionless mass continuity equation for saturation becomes

$$\frac{\partial S(t,y^+)}{\partial t} = \frac{KCq}{\varepsilon\mu_w} \frac{1}{\delta_p{}^2}\left[(p-q-1)S^{p-q-2}\left(\frac{\partial S(t,y^+)}{\partial y^+}\right)^2 + S^{p-q-1}\frac{\partial^2 S(t,y^+)}{\partial y^{+2}}\right]$$
$$- v_f(1\ y^+)\frac{1}{\delta_p}\frac{\partial S(t,y^+)}{\partial y^+} \tag{4.4}$$

Letting $t^+ = \frac{t*|v_f|}{\delta_{f0}}$, the dimensionless equation of the water saturation becomes,

Y. Liu, F. A. Kulacki, *The Effect of Surface Wettability on the Defrost Process*, SpringerBriefs in Applied Sciences and Technology, https://doi.org/10.1007/978-3-030-02616-5_4

$$\frac{\partial S(t^+, y^+)}{\partial t^+} = \frac{\begin{array}{c}\dfrac{\delta_{f0} K C q}{|v_f| \, \epsilon \mu_w \, \delta_p{}^2} \dfrac{1}{} \left[(p - q - 1) S^{p-q-2} + S^{p-q-1} \dfrac{\partial^2 S(t^+, y^+)}{\partial y^{+2}} \right]\\ -\dfrac{\delta_{f0}}{|v_f|} v_f (1 y^+)\\ \times \dfrac{1}{\delta_p} \dfrac{\partial S(t^+, y^+)}{\partial y^+}.\end{array}}{}$$

(4.5)

The boundary and initial conditions are,

$$y^+ = 0, \qquad S = 1 \tag{4.6a}$$

$$y^+ = 1, \qquad S = 0 \tag{4.6b}$$

$$t^+ = 0, \qquad S = 0 \tag{4.6c}$$

The governing equation of water saturation is thus a second order, nonlinear partial differential equation that is to be solved numerically via finite difference Schemes [102–107]. Forward-time and central-space discretization is applied to solve these equations in Eq. (4.5).

Heat transfer in the frost layer is governed by Eq. (3.14). Let $\Theta = \frac{T_f - T_a}{T_i - T_a}$, $y^+ = \frac{y - \delta_p}{\delta_f}$, $U = \frac{\delta_f}{\alpha_f} v_f$, and $t^+ = \frac{\alpha_f t}{\delta_f{}^2}$. The thickness of frost layer and permeation layer varies with time. At a specified time, the energy equation can be simplified as,

$$\frac{\partial \Theta}{\partial t^+} = \frac{\partial^2 \Theta}{\partial y^{+2}} - U \frac{\partial \Theta}{\partial y^+} \tag{4.7}$$

The boundary and initial conditions become,

$$y^+ = 0, \qquad \Theta = 1 \tag{4.8a}$$

$$y^+ = 1, \qquad -\frac{\partial \Theta}{\partial y^+} = Bi\Theta, \quad Bi = \frac{h_a \delta_f}{k_f} \tag{4.8b}$$

$$t^+ = 0, \qquad \Theta = 1 \tag{4.8c}$$

The dimensionless diffusion equation is homogeneous with one nonhomogeneous boundary condition at the permeation–frost interface and one homogeneous boundary condition at the frost–air interface. Thus, the problem comprises a steady state sub-problem with nonhomogeneous boundary conditions, and a transient problem with homogeneous boundary conditions. A forward-time central-space discretization scheme is applied for the numerical solution.

4.2 Accumulation

The water film expands when the permeation layer melts. The temperature in the permeation layer maintains the melting temperature. The water film is expected to be very thin where the Biot number is less than 0.1. A uniform temperature is assumed as

$$T_w = T_s \tag{4.9}$$

The thickness of the water film approximates as

$$\delta_{w(t)} = \frac{q''_s}{\rho_p L_f} t \tag{4.10}$$

4.3 Draining

Conservation of momentum in the water film is described by Eq. (3.22). When the slope angle is 90°, the equation becomes,

$$\frac{\partial u_w}{\partial t} + v_w \frac{\partial u_w}{\partial y} = \frac{\mu_w}{\rho_w} \frac{\partial^2 u_w}{\partial y^2} + g \tag{4.11}$$

Let $u_w^* = \frac{u_w}{U}, v_w^* = \frac{v_w}{U}, t^* = t\frac{U}{\delta_w}, y^* = \frac{y}{\delta_w}$, the dimensionless form of the momentum equation becomes,

$$\frac{\partial u_w^*}{\partial t^*} + v_w^* \frac{\partial u_w^*}{\partial y^*} = \frac{1}{Re} \frac{\partial^2 u_w^*}{\partial (y^*)^2} + \frac{1}{Fr^2} \tag{4.12}$$

where the characteristic velocity $U = |v_f| = \frac{q''_s}{(1-\varepsilon)\rho_i L_f}$, the Reynolds number $Re = \frac{\rho_w U \delta_w}{\mu_w}$, and the Froude number $Fr = \frac{U}{\sqrt{g\delta_w}}$. The Reynolds number is relatively small because the scales of melting velocity and thickness of water film are small, and thus the inertia force is neglected. The momentum equation becomes,

$$\frac{1}{Re} \frac{d^2 u_w^*}{d(y^*)^2} + \frac{1}{Fr^2} = 0. \tag{4.13}$$

A parabolic draining velocity profile is assumed.

$$\frac{du_w^*}{dy^*} = -\frac{Re}{Fr^2} y^* + C_1 \tag{4.14}$$

$$u_w^* = -\frac{Re}{2Fr^2} (y^*)^2 + C_1 y^* + C_2 \tag{4.15}$$

where C_1 and C_2 are constants that can be obtained from the boundary conditions.

Boundary conditions are formulated with regard to surface wettability. For hydrophilic surfaces, the no slip boundary condition is assumed at the solid–water interface. Due to the effect of the boundary layer in the porous medium, a slip velocity is expected at the water–permeation interface.

$$y^* = 0, \qquad u_w^* = 0. \tag{4.16a}$$

$$y^* = 1, \qquad \frac{du_w^*}{dy^*} = \frac{\beta}{\sqrt{K^*}} (u_B^* - \omega^*), \tag{4.16b}$$

where $K^* = \frac{K}{\delta_w^2}$, and $\omega^* = \frac{\omega}{U}$. Solving for C_1 and C_2,

$$C_1 = \frac{Re}{Fr^2} + \frac{\beta}{\sqrt{K^*}} (u_B^* - \omega^*), \tag{4.17a}$$

$$C_2 = 0. \tag{4.17b}$$

The momentum equation for hydrophilic surface is solved with

$$u_w^* = -\frac{Re}{2Fr^2} (y^*)^2 + \left[\frac{Re}{Fr^2} + \frac{\beta}{\sqrt{K^*}} (u_B^* - \omega^*) \right] y^* \tag{4.18}$$

The velocity at water–permeation interface is obtained at $y^* = 1$,

$$u_B^* = \frac{\frac{Re}{2Fr^2} - \frac{\beta}{\sqrt{K^*}} \omega^*}{\left(1 - \frac{\beta}{\sqrt{K^*}} \right)}. \tag{4.19}$$

For the hydrophobic surface, a slip boundary condition is assumed at solid–water interface by Eq. (3.27). Letting $u_s^* = \frac{u_s}{U}, b^* = \frac{b}{\delta_w}$, a dimensionless slip boundary condition is

$$y^* = 0, u_s^* = b^* \frac{du_w^*}{dy^*} \tag{4.20a}$$

The boundary condition at water–permeation interface is described as follows:

$$y^* = 1, \quad \frac{du^*_w}{dy^*} = \frac{\beta}{\sqrt{K^*}}(u^*_B - \omega^*) \tag{4.20b}$$

Solving for the constants C_1 and C_2,

$$C_1 = \frac{Re}{Fr^2} + \frac{\beta}{\sqrt{K^*}}(u^*_B - \omega^*) \tag{4.21a}$$

$$C_2 = b^* C_1 = b^* \left[\frac{Re}{Fr^2} + \frac{\beta}{\sqrt{K^*}}(u^*_B - \omega^*) \right] \tag{4.21b}$$

The momentum equation for the hydrophobic surface is solved with

$$u^*_w = -\frac{Re}{2Fr^2}(y^*)^2 + \left[\frac{Re}{Fr^2} + \frac{\beta}{\sqrt{K^*}}(u^*_B - \omega^*) \right] y^* + b^* \left[\frac{Re}{Fr^2} + \frac{\beta}{\sqrt{K^*}}(u^*_B - \omega^*) \right] \tag{4.22}$$

$$u^*_B = \frac{\left(\frac{1+2b^*}{2} \right) \frac{Re}{Fr^2} - \frac{\beta}{\sqrt{K^*}}(1+b^*)\omega^*}{1 - \frac{\beta}{\sqrt{K^*}}(1+b^*)} \tag{4.23}$$

The average draining velocity and the average draining rate across the water film is

$$u^*_{avg} = \int_0^1 u^*_w dy^*, \tag{4.24}$$

and the average draining rate is

$$\Omega^* = u^*_{avg} \frac{W}{\delta_w}, \tag{4.25}$$

where W is the width of the test plate.

The average draining velocity for the hydrophilic surface is

$$u^*_{avg} = \frac{Re}{3Fr^2} + \frac{\beta}{2\sqrt{K^*}}(u^*_B - \omega^*) \tag{4.26}$$

The average draining velocity for hydrophobic surface is

$$u^*_{avg} = \left(\frac{1}{3} + b^* \right) \frac{Re}{Fr^2} + \left(\frac{1}{2} + b^* \right) \frac{\beta}{\sqrt{K^*}}(u^*_B - \omega^*) \tag{4.27}$$

4.4   Slumping Criteria

The bulk frost is balanced by adhesion force in shear and gravity described by Eq. (3. 33). A ratio is defined corresponding to the competition between these two forces,

$$FR_1 = \frac{G_{f0}}{A_s \tau_i} = \frac{\rho_f g \delta_{f0}}{\tau_i}, \tag{4.28}$$

where density is approximated as $\rho_f = (1 - \varepsilon)\rho_i$.

The tension force in shear of ice decreases with the increase of the surface temperature during the defrost process. To quantify the FR_1, the tension force in shear is assumed to be 100 kPa [67]. Substituting $\varepsilon = 0.5, \rho_i = \frac{900 \text{ kg}}{\text{m}^3}, \delta_f = 3$ mm, FR_1 is

$$FR_1 = \frac{450 \frac{\text{kg}}{\text{m}^3} \times 0.003 \text{ m} \times 9.81 \frac{\text{m}}{\text{s}^2}}{100 \frac{\text{kN}}{\text{m}^2}} = 1.35 \times 10^{-4} \tag{4.29}$$

The body force is much less than the adhesion force in shear and slumping is unlikely to occur when adhesion is much larger than the body force.

During accumulation, water film on hydrophilic surfaces is subjected to gravity, surface tension at the solid–water interface and shear at the water–permeation interface. A ratio is defined as

$$FR_2 = \frac{G_{f0}}{F_{s-w}} = \frac{\rho_f g A_s \delta_{f0}}{\gamma \cdot 2(L + W) \cdot \cos \theta} \tag{4.30}$$

On hydrophobic surfaces, the force ratio is

$$FR_2 = \frac{\rho_f g A_s \delta_{f0}}{\gamma \cdot 2(L + W) \cdot (\cos \theta_{\text{rec}} - \cos \theta_{\text{adv}})} \tag{4.31}$$

Surface tension is defined as the force per unit length. The surface tension of water is 73 mN/m. Substituting $\varepsilon = 0.7, \rho_i = \frac{900 \text{ kg}}{\text{m}^3}, \delta_f = 3$ mm, $L = W = 30$ mm, FR_2 is

$$FR_2 = \frac{(1 - 0.7) \times 900 \frac{\text{kg}}{\text{m}^3} \times (30^2 \times 10^{-6} \times 3 \times 10^{-3}) \text{m}^3 \times 9.81 \frac{\text{m}}{\text{s}^2}}{73 \frac{\text{mN}}{\text{m}} \times (4 \times 30 \times 10^{-3}) \text{ m} \times \cos \theta}$$

$$= \frac{0.8}{\cos \theta} \tag{4.32}$$

The force ratio for hydrophobic surfaces is

$$FR_2 = \frac{0.8}{\cos\theta_{rec} - \cos\theta_{adv}} \tag{4.33}$$

Frost slumping is more likely to occur during the accumulation stage. Surface tension decreases with the surface temperature, and the force ratio increases with time. When the force ratio is greater than 1, gravity outweighs the surface tension, and slumping is likely to occur where a thin water film accumulates on hydrophilic surfaces or where water droplets stick on hydrophobic surfaces.

A part of the meltwater drains at this stage, and the force ratio on the surfaces is time dependent. On hydrophilic surfaces, the force ratio becomes

$$FR_3 = \frac{\rho_f g A_s \delta_{f0} - \rho_w g u_w W \delta_w \Delta t}{\gamma \cdot 2(L + W) \cdot \cos\theta} \tag{4.34}$$

The force ratio on hydrophobic surfaces is

$$FR_3 = \frac{\rho_f g A_s \delta_{f0} - \rho_w g u_w W \delta_w \Delta t}{\gamma \cdot 2(L + W) \cdot (\cos\theta_{rec} - \cos\theta_{adv})} \tag{4.35}$$

The slumping condition in the drainage stage depends on the force competition between the gravity and the surface tension. The bulk mass gets smaller due to the meltwater drainage, while the surface tension decreases with the increase of the surface temperature. The force ratio might be greater than 1 when the decrease due to surface tension is faster than that due to gravity.

A general scaling analysis is applied to define the criterion for the occurrence of slumping during the accumulation stage. As surface tension decreases with increasing temperature, a linear relation is assumed,

$$\gamma = -\frac{\gamma_0}{T_{kp}} T + \gamma_0 \tag{4.36}$$

where γ_0 is the surface tension of water at 0 °C. The force ratio on hydrophilic surfaces becomes,

$$FR = \frac{g\rho_i(1 - \varepsilon)A_s\delta_{f0}}{\left(-\frac{\gamma_0}{T_{kp}}T + \gamma_0\right) \cdot 2(L + W) \cdot \cos\theta} \tag{4.37}$$

The force ratio on hydrophobic surfaces becomes,

$$FR = \frac{g\rho_i(1 - \varepsilon)A_s\delta_{f0}}{\left(-\frac{\gamma_0}{T_{kp}}T + \gamma_0\right) \cdot 2(L + W) \cdot (\cos\theta_{rec} - \cos\theta_{adv})} \tag{4.38}$$

When the aspect ratio is defined as the ratio of plate width to length, $r = W/L$, the force ratio on hydrophilic surfaces becomes,

$$FR = \frac{g\rho_i(1-\varepsilon)W\delta_{f0}}{\left(-\frac{\gamma_0}{T_{kp}}T + \gamma_0\right) \cdot 2(1+r) \cdot \cos\theta}, \tag{4.39}$$

and the force ratio on hydrophobic surfaces becomes,

$$FR = \frac{g\rho_i(1-\varepsilon)W\delta_{f0}}{\left(-\frac{\gamma_0}{T_{kp}}T + \gamma_0\right) \cdot 2(1+r) \cdot (\cos\theta_{rec} - \cos\theta_{adv})} \tag{4.40}$$

Chapter 5
Experimental Design

5.1 Apparatus

The vertical test surface is thermally controlled and mounted on one wall of a chamber in which ambient conditions are controlled (Fig. 5.1). It comprises an acrylic chamber, cooling and heating system for the test surface which is mounted on one vertical wall. The chamber is wrapped with 2.54 cm thick lightweight polystyrene foam insulation (K-factor = 0.26 at 24 °C). A cooling and humidifying system maintains ambient temperature and humidity within the chamber.

Two test surfaces are employed in the experiments: 50 mm × 50 mm and 38 mm × 38 mm. The test surfaces are aluminum and are cooled with a thermo-electric module. Surface temperatures to −20 °C can be maintained during the frost growth period. The hot side of the thermoelectric module is cooled by a cooling water block through which chilled water passes. A film heater is placed between the test surface and the thermoelectric module to provide heating during the defrost process.

The chamber temperature is controlled by the four thermoelectric modules connected in parallel at the top. A fined heat sink is used to absorb heat from the chamber air. Two cooling fans circulate the air inside the chamber at specified volumetric flows of 311 L/min (11 CFM) and 934 L/min (33 CFM). An air pump is used to extract air from the chamber and pipe it back into the chamber after humidifying. A water flask works as a saturator filled with distilled water. Humidity inside the chamber is controlled by a flowmeter connected downstream of the air pump Two high resolution cameras are set up at the front and side of the test chamber to capture the pictures during frost growth and the defrost process. Additional details of the design and construction of the apparatus are given by Liu [107].

© The Author(s), under exclusive license to Springer Nature Switzerland AG 2019   49
Y. Liu, F. A. Kulacki, *The Effect of Surface Wettability on the Defrost Process*,
SpringerBriefs in Applied Sciences and Technology,
https://doi.org/10.1007/978-3-030-02616-5_5

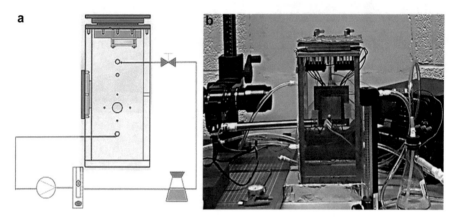

Fig. 5.1 (**a**) Experimental apparatus and (**b**) laboratory setup

Table 5.1 Relation of frost properties with the test conditions

	Time	Humidity	Plate surface temp.	Ambient temp.	Air velocity (*Re*)	Wettability (CA)
Frost density	+	−	+	+	+	−
Frost growth rate	−	+	−	+	+	
Frost thickness	+	+	−	+	+	
Frost mass	+	+	−	+	+	

"+" means dependent variables increase with the increase of the independent variables
"−" means dependent variables decrease with the increase of the independent variables

5.2   Design Analysis

The experiments are designed for the frost and defrost cycles and thus a detailed thermal analysis of the limits of operation of the test surface and chamber are necessary to underwrite data quality and experimental uncertainty.

The test plate temperature can decrease from the melting temperature up to −20 °C. Chamber temperature can vary from the ambient temperature to −10 °C, and chamber humidity can be controlled through a wide range. The temperature variation across the test surface is limited to ~1 °C. During the frost growth period, the surface temperature and the heat flux across the test plate reach to a steady state. During the defrost process, the surface temperature increases with time, and the heat flux across the test plate is variant with time.

Thick frost layers grow with low plate temperature, high ambient temperature, and high humidity. A relation of frost properties with the test condition is shown in Table 5.1. From the table, a dense frost layer is obtained with high plate temperature, high ambient temperature, high air velocity, and low humidity. In the experiments, the chamber temperature is set to the ambient temperature, and the chamber humidity is set ~50%. The plate temperature is set to $-20\ °C$.

The dimensions of the test plate are designed to promote the slumping conditions. When the melting starts, the motion of the frost column is determined by the forces acting on it. Gravity and surface tension are functions of the surface dimensions of the test plate. A static force analysis is described in Chaps. 3 and 4. The ratio of the gravity to the surface tension is

$$FR = \frac{g\rho_i(1-\varepsilon)W\delta_{f0}}{\left(-\frac{\gamma_0}{T_{kp}}T+\gamma_0\right)\cdot 2(1+r)\cdot \cos\theta} \tag{5.1}$$

Substituting with $\cos\theta = 1, \gamma_0 = 76\ \frac{mN}{m}, \varepsilon = 0.6, \delta_f = 1\ mm, r = 1, FR = 1,$ the width of the test plate is 40 mm to achieve the slumping condition.

Chamber Energy Balance

The chamber is a space with controllable temperature and humidity. The temperature can be cooled down by the thermal electric modules on the top of the chamber. During defrosting, heat transfer from the test surface into the chamber causes the ambient temperature to change. Heat transfer into the chamber depends on the heat transfer coefficient on the test surface and the temperature difference between the frost surface and the ambient air inside the chamber,

$$Q_{ch} = h_a A_s \Delta t (T_{fs} - T_{ch}). \tag{5.2}$$

The temperature change inside the chamber is

$$\Delta T_{ch} = \frac{Q_{ch}}{(\rho c_p)_a V_{ch}} \tag{5.3}$$

Heat transfer into the frost layer includes sensible heat and latent heat.

$$Q_{sen} = q''_s A_s \Delta t_{sen} = (mc_p)_f \Delta T_f \tag{5.4}$$

$$Q_{lat} = q''_s A_s \Delta t_{lat} = m_f L_f \tag{5.5}$$

The temperature change inside the chamber becomes:

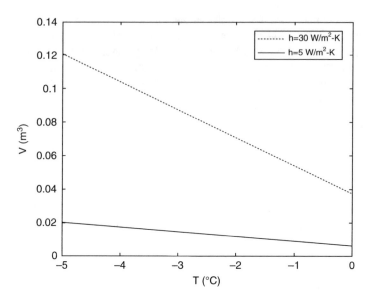

Fig. 5.2 Chamber volume versus chamber temperature for a change of 1 K in the chamber

$$\Delta T_{ch} = \frac{h_a A_s \Delta t_{sen}(T_{fs} - T_{ch}) + h A_s \Delta t_{lat}(T_m - T_{ch})}{(\rho c_p)_a V_{ch}} \tag{5.6}$$

To maintain a constant temperature, chamber volume increases with heat transfer coefficient, test surface area, frost thickness, chamber temperature, and decreases with applied heat flux and porosity. Substituting values with $A_s = 50^2$ mm^2, $T_{fs} = -20$ °C, $h = 5$ W/m^2-K, $\varepsilon = 0.04$, $q''_s = 3000$ W/m^2, $\delta_f = 5$ mm, and $m_f = 6.9$ g, the volume of chamber is a function of the chamber temperature as shown in Fig. 5.2. During the defrost process, the circulation fan stops, and the heat transfer coefficient becomes small. The internal chamber volume is 0.003 m^3.

Heat Transfer Coefficient on the Test Surface and Frost

Heat transfer coefficients differ during frost growth and defrosting. When frost grows, the cooling fan inside the chamber is on to circulate the air. The heat transfer coefficient is calculated with an empirical correlation under the forced convection. During the defrost process, the cooling fan is off, and free convection determines the heat transfer coefficient. When forced convection over the vertical plate dominates, the velocity of circulating air is:

$$u_a = \frac{\text{Volumetric flow rate}}{\text{Area}} = \frac{66 \text{ CFM}}{5'' * 4''} = 2.4 \frac{\text{m}}{\text{s}} \tag{5.7}$$

The Reynolds number is:

$$Re = \frac{\rho u_a L}{\mu_a} = 9000 \text{ (Laminar flow)} \tag{5.8}$$

The film temperature is:

$$T_f = \frac{T_{fs} + T_{ch}}{2} \tag{5.9}$$

and the average Nusselt number for a constant temperature test surface is:

$$\overline{Nu} = \frac{\bar{h}L}{k} = 0.664 Re^{1/2} Pr^{1/3} = 53.2 \tag{5.10}$$

The average heat transfer coefficient is:

$$\bar{h} = \frac{\overline{Nu}k}{L} = 25.5 \text{ W/m}^2\text{K} \tag{5.11}$$

With a constant heat flux at the test surface, local Nusselt number is:

$$Nu_x = \frac{h_x x}{k} = 0.453 Re_x^{1/2} Pr^{1/3} \tag{5.12}$$

And the local heat transfer coefficient is:

$$h_x = \frac{Nu_x k}{x} = \frac{0.453 \left(\frac{\rho u_a x}{\mu_a}\right)^{1/2} Pr^{1/3}}{x} \tag{5.13}$$

The relation of local Nusselt number and local heat transfer coefficient on the test plate is shown in Figs. 5.3 and 5.4.

When circulation fans are off, the heat transfer coefficient for free convection is:

$$\overline{Nu} = \frac{\bar{h}L}{k} = \left(\frac{4}{3}\right)\left(\frac{Gr_L}{4}\right)^{\frac{1}{4}} f(Pr), \tag{5.14}$$

where

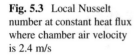

Fig. 5.3 Local Nusselt number at constant heat flux where chamber air velocity is 2.4 m/s

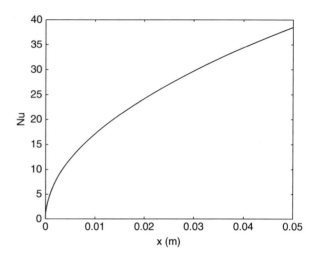

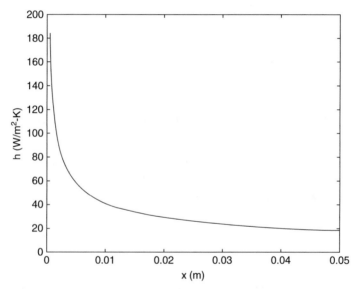

Fig. 5.4 Local heat transfer coefficient at constant heat flux where chamber air velocity is 2.4 m/s

$$f(Pr) = \left(0.75Pr^{1/2}\right)\left(0.609 + 1.221Pr^{1/2} + 1.238Pr\right)^{-1/4} \qquad (5.15)$$

and

Fig. 5.5 Heat transfer
coefficient versus
temperature difference
between the test plate and
surrounding under free
convection where $Pr = 0.7$,
$L = 5.5$ cm

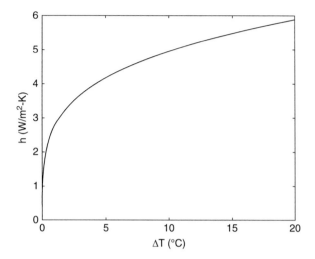

$$Gr = \frac{g\beta(T_{fs} - T_{ch})L^3}{v^2} \tag{5.16}$$

The heat transfer coefficient is a function of the temperature difference between the frost surface and the chamber (Fig. 5.5).

Heat Flux at the Test Surface

During the defrost process, heat flux at the test surface depends on the electric power into the test plate, the melting rate of the frost layer and the heat transfer into the chamber. When assuming a uniform temperature across the test plate, the energy equation before melting can be described as:

$$\left(\rho c_p\right)_{Al} V \frac{dT_{tp}}{dt} + \left(\rho c_p\right)_f \delta_f A_s \frac{dT_f}{dt} = A_s[q''_s - h_a(T_{fs} - T_{ch})] \tag{5.17}$$

When melting starts, air in the frost layer is replaced by meltwater, which forms a permeation layer, and the energy equation becomes:

$$\left(\rho c_p\right)_{Al} V \frac{dT_{tp}}{dt} + \rho_p L_f A_s \frac{d\delta_p}{dt} = A_s[q''_s - h_a(T_{fs} - T_{ch})] \tag{5.18}$$

To investigate the temperature response at the test surface, a simplified equation based on the lumped analysis is applied when Biot number

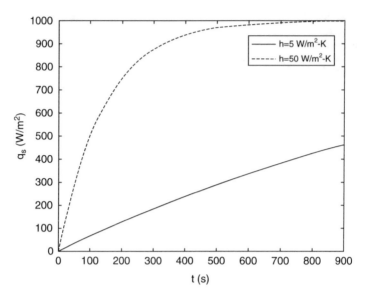

Fig. 5.6 Heat flux at test surface

$$Bi = \frac{h_a \delta}{k} = \frac{50\frac{W}{m^2K}\, 5\text{ mm}}{200\frac{W}{mK}} < 0.1 \tag{5.19}$$

The energy equation for the test plate is:

$$\left(\rho c_p\right)_{Al} V \frac{dT_{tp}}{dt} = A_s[q''_s - h_a(T_s - T_{ch})] \tag{5.20a}$$

$$\left(\rho c_p\right)_{Al} \delta \frac{d\left(T_{tp} - T_{ch}\right)}{dt} = q''_s - h_a(T_s - T_{ch}) \tag{5.20b}$$

The initial temperature at the test surface is:

$$t = 0, \qquad T_s = T_{ch} \tag{5.20c}$$

Heat flux at the test surface is:

$$h_a\left(T_{tp} - T_{ch}\right) = q''_s - [q''_s - h_a(T_s - T_{ch})]e^{-\frac{h_a}{\left(\rho c_p\right)_{Al}\delta}t} \tag{5.21}$$

A relation of heat flux with time is shown for free convection and forced convection in Fig. 5.6. During the defrost process, heat flux at the surface changes with time and is expected to reach a steady value with a large enough heat transfer coefficient.

Heat Sink

The chamber is cooled by thermoelectric modules, and the heat is removed by a heat sink on top of the chamber. The heat load from the chamber is:

$$Q = (\rho c_p)_a V_{ch} \Delta T_{ch} \tag{5.22}$$

Thermal resistance includes fin resistance and base resistance.

$$R_{th,tot} = R_{th,conv} + R_{th,cond} \tag{5.23}$$

$$R_{th,conv} = \frac{1}{h[(N_{fin} - 1)(bl)L + N_{fin}\eta_{fin}2H_{fin}L]} \tag{5.24}$$

$$R_{th,cond} = \frac{H_{tot} - H_{fin}}{k_{Al}WL} \tag{5.25}$$

where N_{fin} is the number of fins, η_{fin} is the fin efficiency, bl is the gap between fins, L is the base length, W is the base width, H_{tot} is the total height of the heat sink, and H_{fin} is the height of the fin. The gap between fins, bl, is:

$$bl = \frac{W - N_{fin}\delta_{fin}}{N_{fin} - 1} \tag{5.26}$$

where δ_{fin} is the thickness of fins. The fin efficiency is given by:

$$\eta_{fin} = \frac{\tanh(mH_{fin})}{mH_{fin}} \tag{5.27}$$

where $m = \frac{2h}{k_{Al}\delta_{fin}}$.

Capacity of the Thermoelectric Module

To achieve $-20\,°C$, the test plate is cooled using a thermoelectric module during the frost growth period. The capacity of the thermoelectric module is verified by using heat balance method. During the frost process, steady state is achieved when the heat flux from the cold side of the thermoelectric module equals the convection heat transfer to the chamber.

$$q''_c = h_c(T_{ch} - T_s) \tag{5.28}$$

The heat transfer coefficient in the chamber is approximated to be 25.5 W/m^2 K. The chamber temperature ranges from -10 to 25 °C, and test surface temperature is from -20 to $0\ ^\circ$C. In order to achieve a large load, the maximum temperature difference and a scale factor of two is applied,

$$q''_c = 2 \times 25.5 \frac{W}{m^2 K}(25 - (-20))K = 2295 \frac{W}{m^2} \tag{5.29}$$

The cold side temperature of the thermoelectric module is:

$$T_c = -\frac{q''_c t}{k_{Al}} + T_s \tag{5.30}$$

As the thermal resistance is relatively small, the cold side temperature of the thermoelectric module is approximated to be the test surface temperature. When assuming the hot side temperature of the thermoelectric module to be 35 °C, the maximum temperature difference becomes 55 °C. Heat transfer on cold side is

$$q_c = q''_c A = 2295 \frac{W}{m^2} \times 0.038^2 m^2 = 3.3\ W \tag{5.31}$$

Referring to the thermoelectric module performance, current is approximated to be 5 A, and voltage is 10 V. The thermoelectric module input power is:

$$Pwr_{in} = I \cdot U = 5A \times 10V = 50\ W \tag{5.32}$$

Heat transfer on hot side is:

$$q_h = Pwr_{in} + q_c = 53.3\ W \tag{5.33}$$

The hot side temperature of the thermoelectric module is:

$$T_h = T_a + R_{th,\,fin} q_h \tag{5.34}$$

The thermal resistance of the heat sink is approximated with 0.1 °C/W. The hot side temperature is:

$$T_h = 25\ ^\circ C + 0.1 \frac{^\circ C}{W} \times 53.3W = 30\ ^\circ C \tag{5.35}$$

As the calculated hot side temperature is smaller than the assumed hot side temperature of 35 °C, the thermoelectric module is verified to be able to cool the test surface down to $-20\ ^\circ$C.

5.3   Data Acquisition and Uncertainty Estimates

Two important variables in the experiments are defrost time and defrost efficiency. A set of experiments carry out on different test surfaces with varying wettability. In Chap. 3, a theoretical efficiency is defined as the ratio of the energy required to melt the frost to the total heat applied at the surface.

$$\eta_{df} = \frac{(1 - \varepsilon)\rho_i \delta_f \left[L_f + (c_p)_i \Delta T_f\right]}{q''_s t_{df}} \tag{3.32}$$

A practical definition of defrost efficiency is

$$\eta_{df} = \frac{\text{energy required to melt the frost layer}}{\text{total actual energy consumption}} \tag{5.36}$$

$$\eta_{df} = \frac{m_f \left(L_f + c_{p,f}\Delta T_f\right)}{\text{Pwr} \times t_{df}} \tag{5.37}$$

The total energy required includes the sensible and latent heat to melt the frost layer. The energy consumption, Pwr, is the power input multiplied by the total defrost time. Thus defrost time includes the time for surface preheating, frost melting and the removal of retained water by drying the surface.

The experimental data sets include a number of measurement variables and dependent variables. During the frost formation period, measurement variables are the temperature and humidity of the chamber, saturated air flow, temperature of the test surface, heat flux across the test plate, frost growth time, and power input. For frost properties, when measuring frost thickness and mass directly, frost density is a dependent variable. During the defrost process, measurement variables are the ambient temperature and humidity in the chamber, test surface temperature, heat flux at the test surface, air flow, power input, defrost time, frost thickness, and frost mass. Defrost efficiency is a dependent variable.

The ambient temperature inside the chamber is measured by a shielded type T thermocouples installed at the chamber wall and a thin-film 100 Ω platinum RTD included in the relative humidity temperature transmitter. The humidity transmitter with a current output 4–20 mA is used to monitor the relative humidity inside the chamber. Saturated air flow is adjusted by a glass tube flowmeter to control the chamber humidity.

The temperature of the test surface is measured with five type T thermocouples. Four of them are placed into the holes at the four corners of the test plate, and one is in the groove at the center. Thermocouples are sealed with high thermal conductive epoxy. Heat flux across the test surface is measured by a heat flux sensor that functions as a self-generating thermopile transducer. The readout is obtained by connecting to a DC micro-voltmeter. A built-in Type-K thermocouple is used to measure the heat flux sensor temperature. When the test plate is dry, the heat transfer coefficient is calculated by measuring the heat flux across the test surface, the temperature at the test surface, and the air temperature inside the chamber at steady state.

Frost thickness is measured via imaging. A camera is seated at the side of the test chamber to capture images of the frost thickness (Fig. 5.1). The thickness measured by imaging might be larger than the actual thickness of the frost layer due to the edge effects. A correlation derived by Janssen [10] is used in the frost thickness measurement.

$$\delta_f = 0.0045 \Theta^{(\Theta - 4.721)} t^{\Theta}, \tag{5.38}$$

where the dimensionless temperature is

$$\Theta = \frac{T_a - T_{dp}}{T_a - T_w}. \tag{5.39}$$

Total mass of the frost column is calculated by weighing the saturator before and after the frost growth. The mass of the meltwater is also measured. Frost mass can be determined by obtaining the weight reduction of the water in the saturator and the change of vapor pressure in the chamber, or by weighing the meltwater.

$$\Delta m_f = \Delta m_{sat} - \Delta m_v \tag{5.40}$$

$$\Delta m_v = \frac{\Delta P_v V M_v}{\bar{R} T_{ch}} \tag{5.41}$$

$$\Delta P_v = \Delta \phi P_g \tag{5.42}$$

where ϕ is the relative humidity of air inside the chamber, and P_g is the saturated pressure at the chamber temperature.

Uncertainty Estimates

The uncertainty of the direct measurement variables refers to the manufacturer's specification as listed in Table 5.2.

The uncertainty of dependent variables is determined by the root sum of squares method. Uncertainty of frost thickness depends on the pixel calibration by using the visual imaging technique. An uncertainty of ± 0.01 mm can be obtained with good calibration. The uncertainty of the dependent variables is listed in Table 5.3.

Table 5.2 Instrument accuracies

Measurement	Instrument	Accuracy
Temperature	Type-T thermocouple	±0.5 °C
Humidity/Temperature	Omega HX94C	±2%/±0.6 °C
Heat flux	Omega HFS-4	±0.5%
Flow	Omega FL2012	±5%
Time	Agilent BenchLink Data Logger 3	0.01 s
Mass	Electrical scale	±3 mg

Table 5.3 Uncertainty of dependent variables

Measurement	Accuracy
Frost thickness	±0.01 mm
Frost mass	±5%
Defrost efficiency	±10%

Table 5.4 Characterization of the sample surfaces

Surface structure	Sample 1	Sample 2	Sample 3	Sample 4	Sample 5
Surface material	Al 6061	Al 6061	Al 6061	Al 6061	Al 6061
Surface machining	32 μin. or 0.8 μm	32 μin. or 0.8 μm	32 μin. or 0.8 μm	32 μin. or 0.8 μm	32 μin. or 0.8 μm
Surface finish with 320 grit sandpaper	8 μin. or 0.2 μm	8 μin. or 0.2 μm	8 μin. or 0.2 μm	8 μin. or 0.2 μm	8 μin. or 0.2 μm
Surface coating	Hydrochloric acid	Commercial hydrophilic coating	Plain	Solution of stearic acid and acetone	Commercial hydrophobic coating
Static contact angle (°)	<5	51	71	119	146

5.4   Surface Preparation and Characterization

The aluminum test surfaces are treated for five different contact angles. Surface wettability can be modified in two ways. One way is to change the microstructure of the surface, and another is to alter surface energy with chemical treatment. The treatment methods are characterized in Table 5.4. The superhydrophilic surface is obtained by soaking the test plate in 17% hydrochloric acid solution for about 10 min and rinsing with distilled water.

A hydrophilic coating applies to the test plate for the hydrophilic surface with a contact angle of 51°. The hydrophobic surface is obtained by soaking the test plate in a solution of stearic acid and acetone (0.284 g/100 ml). A commercial hydrophobic coating, NeverWet® superhydrophobic coating, applies on the plate to get a superhydrophobic surface with a contact angle of about 146°.

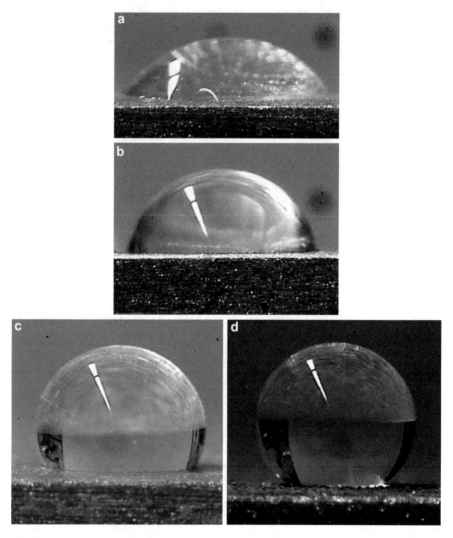

Fig. 5.7 Static contact angles. (**a**) CA = 51°; (**b**) CA = 71°; (**c**) CA = 119°; (**d**) CA = 146°

Static contact angles are measured using sessile drop method. A 1 ml needle pump is used to apply droplet on the test surface with one droplet volume of about 3 µl. The geometries of the droplets are taken with Bigcatch DCM510C and Navitar zoom 7000 as shown in Fig. 5.7. The contact angles are measured using the open source computer program ImageJ.[1] The advancing and receding contact angles are measured on a tilted plate with adjustable sliding angles. The pictures are taken

---

[1] ImageJ is a public domain, Java-based image processing program developed at the US National Institutes of Health.

Fig. 5.8 Dynamic contact angles. $\Theta_{rec} = 107°$, $\theta_{adv} = 149°$ (sliding angle ~28°)

where the droplet grows large enough to roll down the tilted plate as shown in Fig. 5.8.

Chapter 6
Results

6.1 Analytical and Numerical Results

In the absorption of meltwater, frost at the solid surface melts, and the melting rate depends on the heat flux applied at the solid surface, latent heat of fusion, and density of the permeation layer. The relation of the melting rate to the heat flux and the porosity of the frost layer is described in Eq. (3.3) and shown in Figs. 6.1 and 6.2. The rate of melting increases with the porosity, and when the frost layer is less dense, the melting process goes faster. The melting rate increases from 0.01 to 0.1 mm/s when the porosity increases from 0.1 to 0.9. When the porosity is 0.4, the rate of melting is 0.016 mm/s. For a frost layer with a thickness of 3 mm, the melting time is 187 s. The dependence of the melting rate on the heat flux at the solid surface is linear, directly proportional to the heat flux. The melting rate increases from 5.5×10^{-4} to 0.03 mm/s when the heat flux increases from 300 to 6000 W/m^2 when the porosity is 0.4.

In the absorption of meltwater, saturation, S, is a function of time and location as described in Eqs. (3.10) and (4.4). The distribution of saturation depends on the permeability of the porous medium and capillary pressure, which are functions of saturation. The relation of the water saturation with time and location are shown in Figs. 6.3 and 6.4. The permeability power index and the capillary pressure are not available for the frost column in the current literature, and therefore the values used herein are for snow [76–78]. It is shown that saturation increases with time and decreases with location due to the meltwater transport. At the location that is close to the surface, saturation increases rapidly as the boundary condition at the surface is assumed to be $S = 1$.

The energy balance in frost layer is described in Eqs. (3.12) and (4.7). The dimensionless temperature distribution and evolution are shown in Figs. 6.5 and 6.6. Dimensionless temperature decreases with time and location. Dimensionless

© The Author(s), under exclusive license to Springer Nature Switzerland AG 2019
Y. Liu, F. A. Kulacki, *The Effect of Surface Wettability on the Defrost Process*,
SpringerBriefs in Applied Sciences and Technology,
https://doi.org/10.1007/978-3-030-02616-5_6

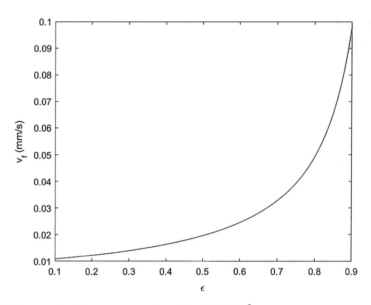

Fig. 6.1 The rate of melting versus porosity. $q_s'' = 3000$ W/m^2

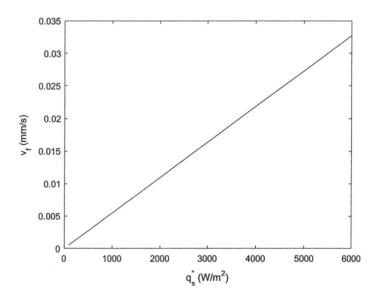

Fig. 6.2 The rate of melting versus heat flux applied at the test surface. $\varepsilon = 0.4$

temperature at the frost–air interfaces is close to 0.14 at 10 s. The dimensionless temperature profile becomes flat at about the half of the thickness.

Draining velocities are described in Eqs. (4.31) and (4.33). The velocity profiles are determined by the boundary conditions at the solid–water interface and at the water–permeation interface. The curves in Fig. 6.7 are the dimensionless drainage

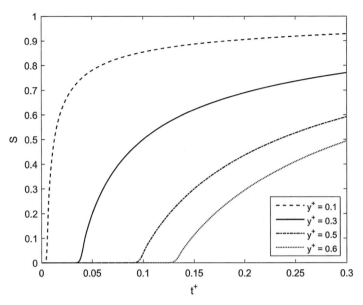

Fig. 6.3 Water saturation versus time. $q_s'' = 3000$ W/m^2, $\varepsilon = 0.6$, $\delta_f = 5$ mm, $p = 3$, $q = 1$, $K = 3 \times 10^{-12}$ m^2, $C = 43$ Pa

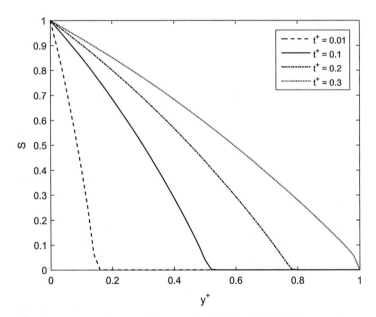

Fig. 6.4 Water saturation versus location. $q_s'' = 3000$ W/m^2, $\varepsilon = 0.6$, $\delta_f = 5$ mm, $p = 3$, $q = 1$, $K = 3 \times 10^{-12}$ m^2, $C = 43$ Pa

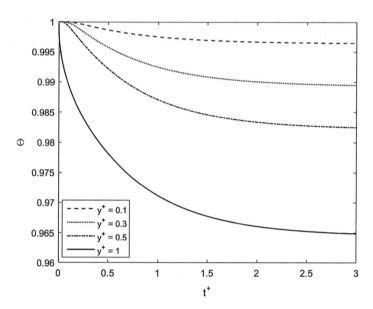

Fig. 6.5 Dimensionless temperature of frost layer versus time. $q_s'' = 3000$ W/m^2, $\varepsilon = 0.7$, $\delta_f = 5$ mm, $h_a = 5$ W/m^2-K

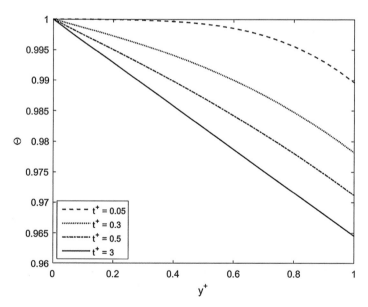

Fig. 6.6 Dimensionless temperature distribution in frost layer. $q_s'' = 3000$ W/m2, $\varepsilon = 0.7$, $\delta_f = 5$ mm, $h_a = 5$ W/m^2-K

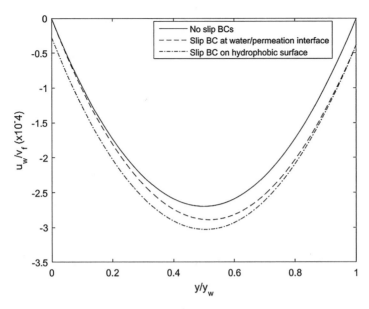

Fig. 6.7 Comparison of draining velocities with different boundary conditions. $\alpha = 1.2$, $K = 10^{-9}$ m^2, $b = 20$ μm

Table 6.1 Comparison of drainage velocity and average drainage rate

	Maximum velocity (m/s)	Average velocity (m/s)	Average drainage rate (m³/s)	Improvement of draining rate (%)
No slip condition at the solid–water interface and at the water–permeation interface	0.437	0.288	8.9×10^{-6}	
No slip condition at the solid–water interface and slip condition at the water–permeation interface	0.468	0.318	9.83×10^{-6}	10
Slip condition at the solid–water interface and at the water–permeation interface	0.49	0.341	1.05×10^{-5}	7.1

Note: Melting rate is 0.016 mm/s where the heat flux applied is 3000 W/m^2 and the porosity is 0.4
The thickness of water film is 0.8 mm and the width of the plate is 38 mm

velocities with respect to the various boundary conditions. Three boundary conditions are presented, and they are no slip boundary conditions at the solid–water interface and at the water/permeation interface, no slip boundary condition at the solid–water interface, and slip boundary condition at the water–permeation interface, slip condition for hydrophobic surface at the solid–water interface and slip condition at the water–permeation interface. The maximum draining velocity occurs when slip conditions are applied to the boundaries at the two interfaces. To better show the

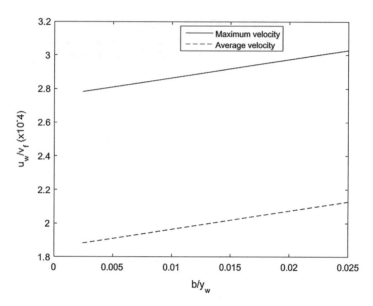

Fig. 6.8 The effect of slip length on the draining velocity

improvement of drainage process on the hydrophobic surface, the maximum and the average drainage velocity and the average drainage rate are quantified and listed in Table 6.1. The drainage velocity depends on fluid properties, the slip length at the solid–water interface, the water film thickness, the slip coefficient at the water–permeation interface, and the permeability of the porous medium.

Slip conditions result from an interplay of many physical and chemical parameters. The influencing factors include the surface wettability, surface roughness, and impurities and shear rate. On the macroscopic level, slip length can be explained as a formation of gas film or phase separated lubricant with lower viscosity between the fluid and the solid wall. An engineered nanostructured superhydrophobic surface can minimize the liquid–solid contact area so that the liquid flows over a layer of air. The surface demonstrates dramatic slip effects with a slip length of ~20 μm in water flow. The effect of slip length on the drainage velocity is shown in Fig. 6.8.

The slip coefficient is experimentally determined and depends on the structure of the material at the surface. Experiments show that the slip coefficient of a metal material depends directly on the average pore diameter at the interface. The study on the slip coefficient applies to the experimental condition where the gap between the permeable wall and the impermeable wall is much larger than the pore size of the porous material. Taylor and Richardson [91, 92] propose a mathematical model and compare the results between the theory and the experiments. The study demonstrates that the slip coefficient is not independent of the external means that producing the external tangential stress. Sahraoui and Kaviany [93] investigate the two-dimensional flow field near the porous medium made of cylinders. The simulation shows that the slip coefficient depends on structure, porosity, flow direction,

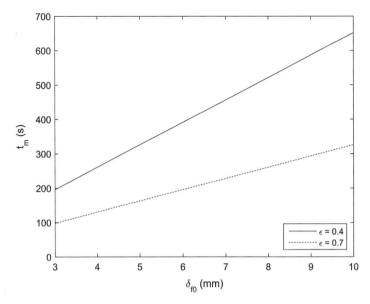

Fig. 6.9 Melting time versus the initial frost thickness. $q_s'' = 3000$ W/m^2, $T_a = -5$ °C

Reynolds number, the extent of the plain medium, and the nonuniformities in the arrangement of surface particles.

During the drainage stage, defrost time and efficiency are given in Eqs. (3.31) and (3.32). Defrost time and efficiency are determined by the heat flux applied at the test surface, the thickness of the water film, and the porosity and thickness of the frost layer. Defrost time consists of melting time and draining time. Porosity and applied heat flux are the two factors that can influence melting time significantly. The relation of melting time to frost thickness and ambient temperature are shown in Figs. 6.9 and 6.10.

Draining time depends on porosity, frost thickness, the height of the plate and water film thickness, as shown in Figs. 6.11, 6.12, 6.13, and 6.14. It is shown that water film thickness significantly influences drainage time. Draining time decreases with increase of porosity and increases with frost thickness. Draining time decreases from 137 to 0.14 s when water film thickness increases from 0.1 to 1 mm, and it decreases sharply with water film thickness from 0.1 to 0.2 mm and goes flat around 0.4 mm.

For relatively large thicknesses of the water film, draining time is small compared to melting time. The relation of defrost time with porosity, water film thickness, heat flux, and height of the test plate are shown in Figs. 6.15, 6.16, 6.17, and 6.18. Defrost time is sensitive to the water film thickness and the height of the test plate. The influence of water film thickness on the defrost time is illustrated in Figs. 6.15 and 6.16. The variation between the melting time and the draining time is relatively small where the water film thickness equals 0.3 mm. Defrost time is thus determined by the

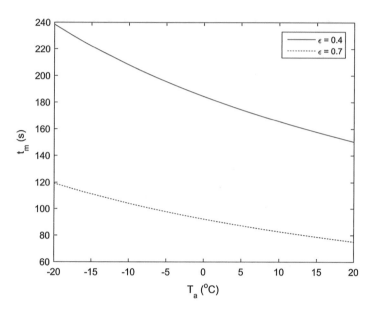

Fig. 6.10 Melting time versus ambient temperature. $q_s'' = 3000$ W/m^2, $\delta_{f0} = 3$ mm

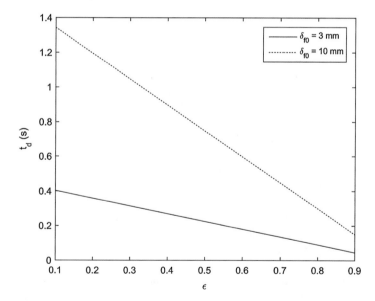

Fig. 6.11 The relation of draining time with porosity. $H = 38$ mm

melting time and the draining time. When the water film thickness equals 0.6 mm, defrost time is mainly determined by the melting time, and the draining time is relatively small. The influence of the height of test plate on defrost time is shown in

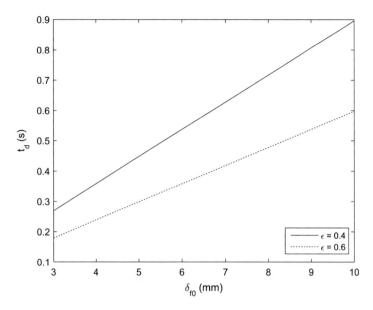

Fig. 6.12   The relation of draining time with the initial frost thickness. $H = 38$ mm

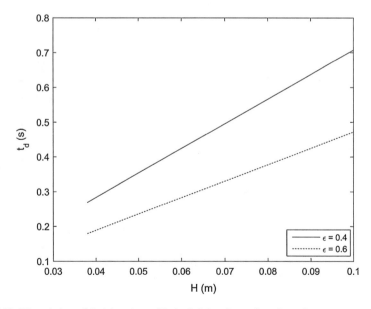

Fig. 6.13   The relation of draining time with the height of test plate. $\delta_{f0} = 3$ mm

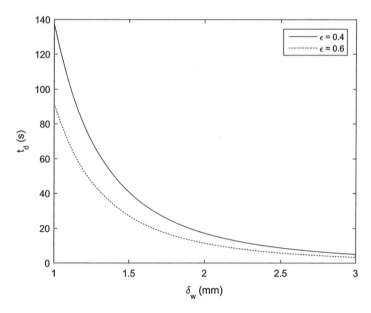

Fig. 6.14 The relation of draining time with the water film thickness. $H = 38$ mm

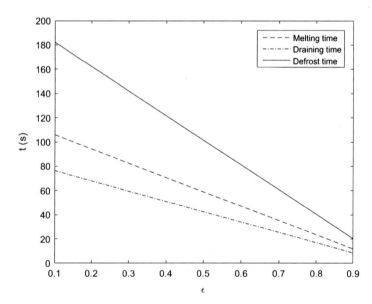

Fig. 6.15 The relation of defrost time with porosity. $q_s'' = 8000$ W/m^2. $H = 380$ mm. $\delta_w = 0.3$ mm

Fig. 6.17. When the height of test plate equals 38 mm, draining time is relatively small, and the defrost time is mainly determined by the melting time. Defrost time increases with the decrease of the heat flux as shown in Fig. 6.18.

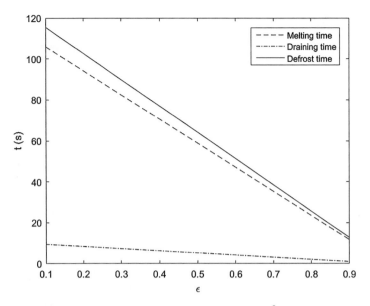

Fig. 6.16 The relation of defrost time with porosity. $q_s'' = 8000$ W/m^2. $H = 380$ mm. $\delta_w = 0.6$ mm

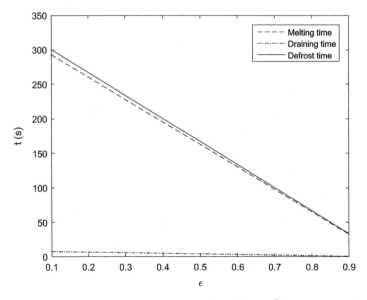

Fig. 6.17 The relation of defrost time with porosity. $q_s'' = 3000$ W/m^2. $H = 38$ mm. $\delta_w = 0.3$ mm

The relation of draining time and defrost time to surface wettability is shown in Figs. 6.19 and 6.20. When the height of the test plate is relatively small and the water film thickness is relatively large, the melting time is much larger than the draining time. Defrost time is mainly determined by the melting time. In this scenario, the

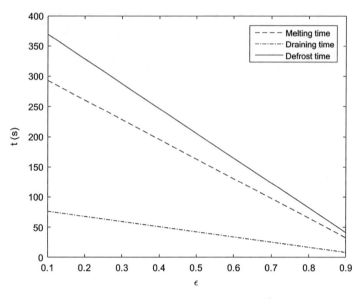

Fig. 6.18 The relation of defrost time with porosity. $q_s'' = 3000$ W/m^2. $H = 380$ mm. $\delta_w = 0.3$ mm

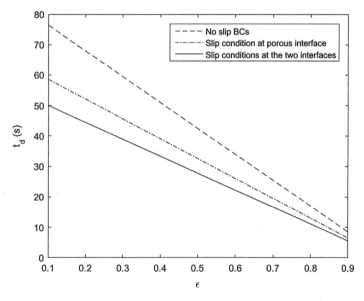

Fig. 6.19 Comparison of the draining time with different boundary conditions. $H = 380$ mm. $\delta_w = 0.3$ mm

effect of surface wettability on the defrost time is insignificant. The smallest drainage time occurs when slip boundary conditions apply at both the solid–water interface and the water/permeation interface. When the water film gets thinner, the effect of

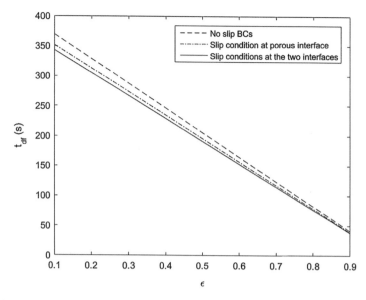

Fig. 6.20 Comparison of defrost time with different boundary conditions. $q_s'' = 3000$ W/m^2. $H = 380$ mm. $\delta_w = 0.3$ mm

the slip boundary conditions gets larger on the average velocity and the draining time. Defrost time is smallest with slip conditions at both interfaces. The effect of the slip conditions is less significant on defrost time compared to that on the draining time.

Defrost efficiency depends on applied heat flux at the solid surface, water film thickness, height of the test plate, heat transfer coefficient at the permeation–air interface, ambient temperature, and slip conditions at the solid–water interface and at the water–permeation interface. For the given conditions, defrost efficiency with slip condition is improved about 5.5% compared to that with no slip boundary condition at the solid–water interface.

Slumping Criterion

Slumping conditions have been discussed in each stage of the melting process. A general formula is defined based on the static force analysis as described in Eqs. (4.39) and (4.40). The ratio of gravity to the retaining force is defined as a measure of the slumping possibilities. Retaining forces vary with the different stages in the melting process. In the absorption stage, the retaining force is the adhesion due to shear force. In the accumulation and the drainage stages, the retaining forces are surface tension around the perimeters of the frost column. The ratio of gravity to surface tension is bigger in the accumulation stage. The possibility of frost slumping

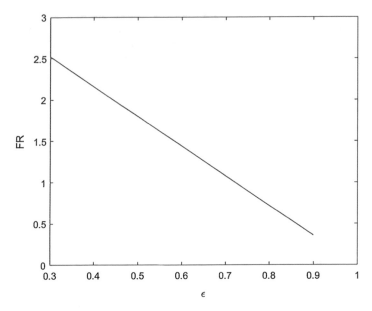

Fig. 6.21 Force ratio versus porosity. $W = 38$ mm. $\delta_{f0} = 3$ mm. $r = 1$. CA $= 20°$

is thus higher during the accumulation stage. The force ratio is related to the porosity and thickness of the frost layer, the geometry of the test plate, the temperature at the test surface, and the surface characteristics of the test surfaces.

The force ratio with respect to porosity is shown in Fig. 6.21. The slumping ratio decreases from 2.5 to 0.36 when porosity ranges from 0.3 to 0.9. When porosity is larger than 0.7, the frost layer is very loose, and slumping is unlikely to occur, based on estimates using the given parameters. Dense frost falls off more easily compared to the loose frost layer. The force ratio with respect to the aspect ratio is shown in Fig. 6.22. At the given width of the test plate, the slumping force ratio decreases from 2.6 to 0.7 when the aspect ratio increases from 0.1 to 3. When the aspect ratio is greater than 1.8, the slumping force ratio is less than 1. A larger force ratio favors slumping at a given width of the test plate. Temperature has influence on the surface tension. The higher the temperature, the smaller the surface tension is. The force ratio thus is related to temperature as shown in Fig. 6.23. The force ratio increases from 0.97 to 1.07 when the temperature increases from -5 °C to 30 °C. In the absorption stage, the temperature at the test plate surface maintains the melting temperature as the mixture of the ice crystals and the meltwater is attached to the test plate.

The effect of surface wettability on the force ratio depends on the physical models of the slumping condition. A thin water film is assumed to form on hydrophilic surfaces during the melting process, and the surface tension depends on the contact angle of water. For hydrophobic surfaces, water droplets are assumed to distribute on the test plate, and the surface tension depends on the receding and advancing contact angles.

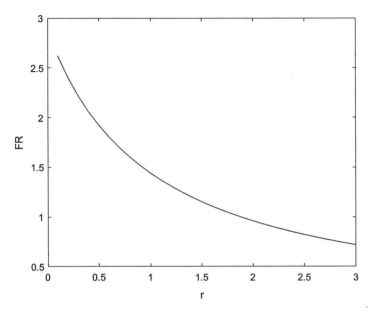

Fig. 6.22 Force ratio versus aspect ratio of the test plate. $W = 38$ mm. $\delta_{f0} = 3$ mm. $\varepsilon = 0.6$. $CA = 20°$

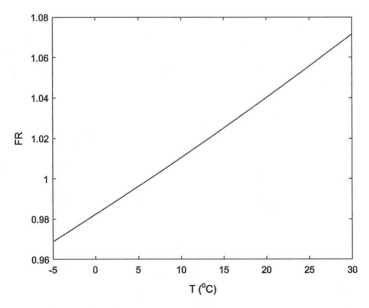

Fig. 6.23 Force ratio versus temperature at the surface. $W = 38$ mm. $\delta_{f0} = 3$ mm. $r = 1.2$, $\varepsilon = 0.6$. $CA = 20°$

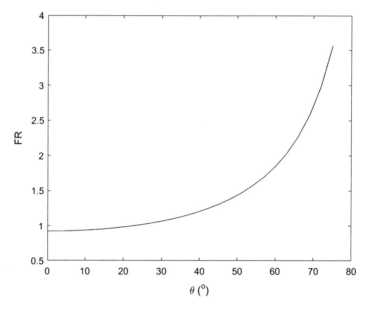

Fig. 6.24 Force ratio versus contact angle on hydrophilic surfaces. $W = 38$ mm. $\delta_{f0} = 3$ mm. $r = 1.2$, $\varepsilon = 0.7$

The force ratio with respect to the contact angle on hydrophilic surfaces is shown in Fig. 6.24. When the contact angle is less than 50°, the force ratio increases smoothly with the contact angle. The force ratio with respect to the receding contact angle and the contact angle hysteresis for hydrophobic surfaces is shown in Fig. 6.25. Three contact angle hysteresis values are applied, and they are 5°, 15°, and 30°. The force ratio increases with the receding contact angle and decreases with the contact angle hysteresis. When the contact angle hysteresis is 5°, the force ratio is about five times of that when the contact angle hysteresis is 30°. The slumping condition is favorable with a large force ratio where the surface energy is low and the contact angle hysteresis is small.

Based on the assumptions on the physical mechanism of slumping condition, the force ratio is the smallest where the surface is fully wetted. The force ratio is much greater than unity on hydrophobic surfaces where the contact angle hysteresis is 5°. With the increase of the contact angle hysteresis, the difference of the force ratio on hydrophobic and hydrophilic surfaces decreases. The slumping would be likely to occur on hydrophobic surfaces with small contact angle hysteresis. With the specified surface temperature, the relation of slumping ratio with respect to porosity, aspect ratio and surface contact angle are shown in Figs. 6.26 and 6.27.

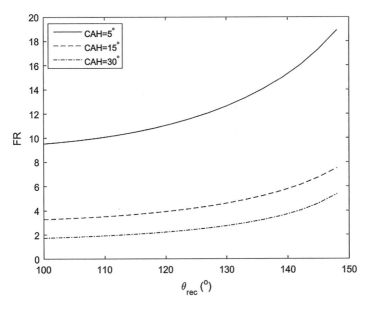

Fig. 6.25 Force ratio versus receding contact angle on hydrophobic surfaces. $W = 38$ mm. $\delta_{f0} = 3$ mm. $r = 1.5$, $\varepsilon = 0.7$

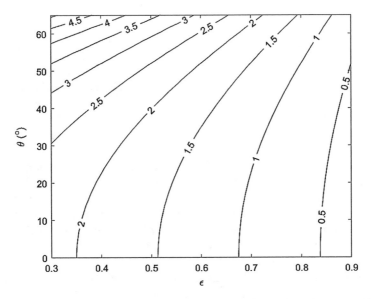

Fig. 6.26 Force ratio versus porosity and contact angle. $W = 38$ mm. $\delta_{f0} = 3$ mm. $r = 1.2$

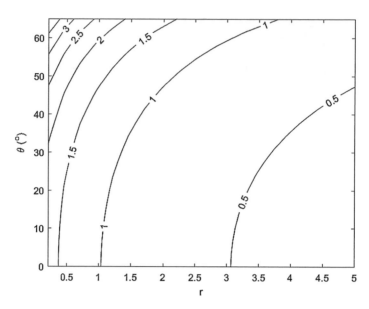

Fig. 6.27 Force ratio versus aspect ratio and contact angle. $W = 38$ mm. $\delta_{f0} = 3$ mm. $\varepsilon = 0.7$

6.2   Experimental Results

Experiments are run for two test surfaces: 50 mm × 50 mm and 38 mm × 38 mm surface. The larger surface area is tested to reduce the edge effects. Surface contact angles are 5°, 71°, and 146° (Table 6.2). Test conditions for frost growth are shown in Table 6.3. The test surface is cooled to -19 °C and the ambient chamber temperature is maintained at 19 °C to achieve high supercooling. A relative humidity of as high as 0.51 is used to increase supersaturation for frost growth.

Normal and in plane views of the frost formation process for the superhydrophilic, plain, and superhydrophobic surfaces are shown in Figs. 6.28, 6.29, 6.30, 6.31, 6.32, and 6.33. The superhydrophilic and surperhydrophobic surfaces are relatively rough compared to the plain surface. At the first 10 min, the photos record the vapor nucleation and the frost formation processes. Condensation is observed over 5 min on all the different surfaces. Observations show that the superhydrophobic surface does not significantly retard nucleation and frost formation. It can be seen that condensation spreads over the test surface in 5 min from Fig. 6.32b. The reason might be that high supercooling degree and high relative humidity speed up the vapor nucleation process in the experiments. Crystal growth starts at ~10 min, followed by the frost growth at all the different surfaces. Frost gets denser where part of the vapor diffuses into it and gets thicker with the deposition of the vapor on the frost surface. The densification and thickening processes are recorded in the photos taken in the middle and at the end of the frost growth process where $t = 1$, 3 and 6 h.

Table 6.2 Test samples

Test plate identifier	Test plate dimensions (mm)	Test surface CA (°)
TestID-1	38 × 38 × 3.8	<5
TestID-2	38 × 38 × 3.8	~71
TestID-3	38 × 38 × 3.8	~146
TestID-4	50 × 50 × 5	<5
TestID-5	50 × 50 × 5	~71
TestID-6	50 × 50 × 5	~146

Table 6.3 Frost formation conditions

	Surface temperature (°C)	Chamber temperature (°C)	Chamber relative humidity (%)	Dew point (°C)	Supercooling (°C)
Test surface 38 × 38 mm	−18 ± 2	19 ± 1	46 ± 5	7	25
Test surface 50 × 50 mm	−16 ± 2	19 ± 1	46 ± 5	7	23

Frost properties of the six test surfaces are listed in Table 6.4. The frost mass increases with time, as shown in Fig. 6.34. The density increases with time, but the variation of the density on the different surfaces is not significant. The reason could be that the temperature at the test surface and the temperature and humidity inside the chamber change slightly, which would influence the thickness accordingly. The uncertainties of the thermocouples and the humidity transmitters also change the dew point temperature and, thus, the thickness to some extent. Also, due to edge effects, the frost on the edges are thicker and denser compared to frost on the center of the surface.

The defrost processes were run on the six test surfaces for superhydrophilic, plain, and superhydrophobic surfaces, as shown in Figs. 6.35, 6.36, and 6.37. The physical mechanisms of the melting process are very different on the surfaces with different degrees of wettability. On the superhydrophilic surface and the plain surfaces, frost melts first at the center where frost is less dense and thicker. A large portion of the meltwater is absorbed into the frost layer at the sides where the frost is denser and thicker. The remaining meltwater takes the form of water film on the superhydrophilic surface and retention droplets on the plain surface. The frost at bottom then melts and drains on superhydrophilic surface. With the water accumulation at the side surfaces, part of the meltwater drains along the sides. The remaining frost falls off with the meltwater. When all the frost on the front surface melts away, a water film band is observed at the bottom part of the superhydrophilic surface, and nonuniform retention droplets are left on the plain surface.

The melting process on the superhydrophobic surface varies greatly from that on the superhydrophilic and plain surfaces. Frost slumping occurs on the test surfaces where frost grows for 3, 4, and 6 h with high heating power and lower heating power

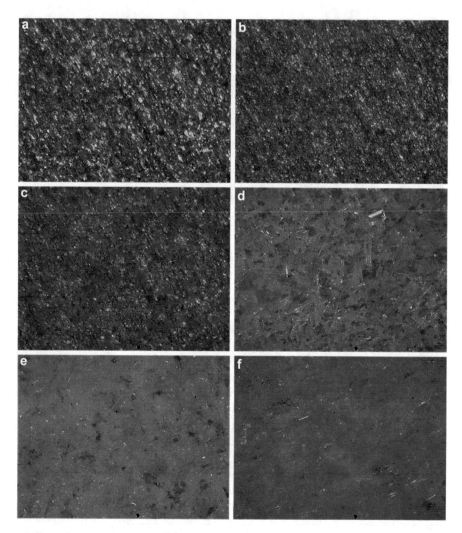

Fig. 6.28 Front view of the frost formation at test surface (TestID-1). (**a**) $t = 0$; (**b**) $t = 5$ min; (**c**) $t = 10$ min; (**d**) $t = 1$ h; (**e**) $t = 3$ h; (**f**) $t = 6$ h

applied separately. When heating applies on the superhydrophobic surface for a period of time, the frost at the center seems to detach from the surface first, and then the frost falls off as a whole piece with the heating continues. In the slumping process, the frost layer looks like a rigid mass, and no water drainage is observed. After the frost slumping, no droplet is left on the test surface. The different melting mechanisms might be explained by the two defrost models by Sanders [56]. The absorption model applies to the superhydrophilic and the plain surfaces. In the melting process, meltwater is absorbed into the frost layer, and the frost layer maintains the contact with the test surface. Different from the absorption model,

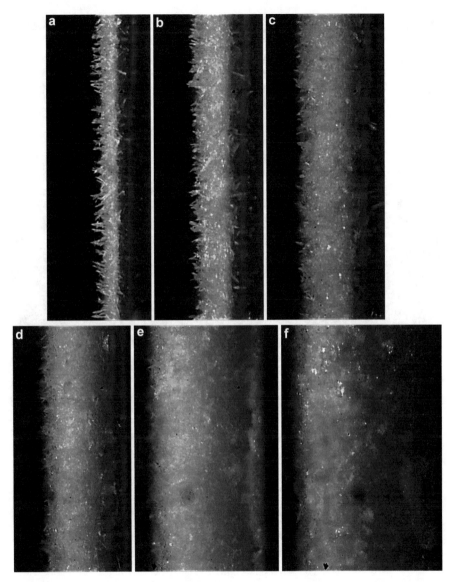

Fig. 6.29 Side view of the frost formation at test surface (TestID-1). (**a**) $t = 10$ min; (**b**) $t = 20$ min; (**c**) $t = 30$ min; (**d**) $t = 1$ h; (**e**) $t = 3$ h; (**f**) $t = 6$ h

an air gap model applies to the superhydrophobic surface. During the melting process, meltwater is absorbed into the frost layer leaving an air gap at the solid–frost interface. The air gap increases the thermal resistance and reduces the heat transferred into the frost layer. The frost column drops off at the point where the gravity outweighs the adhesion force.

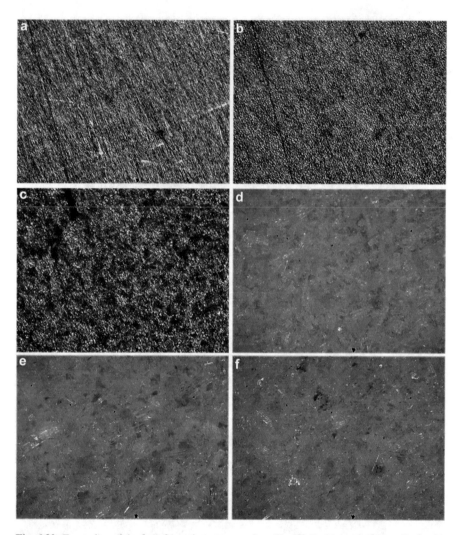

Fig. 6.30 Front view of the frost formation at test surface (TestID-2). (**a**) $t = 0$; (**b**) $t = 5$ min; (**c**) $t = 10$ min; (**d**) $t = 1$ h; (**e**) $t = 3$ h; (**f**) $t = 6$ h

The small test plates (38 mm $\times$ 38 mm) are heated by the thermoelectric module with the DC power settings of 60, 15, and 6 W. The defrost properties are listed in Table 6.5. During the defrost process, the actual output of the DC power varies with time. Heat flux at the test plate varies with time as shown in Fig. 6.38. Heat flux increases rapidly at the first 20 s because the temperature at the heat flux sensor increases much quicker than that at the test plate. When the temperature difference between the heat flux sensor and the test plate becomes smaller, the heat flux decreases accordingly. The profile of the heat flux is consistent with the profile of the temperature difference between the heat flux sensor and the test plate as shown in

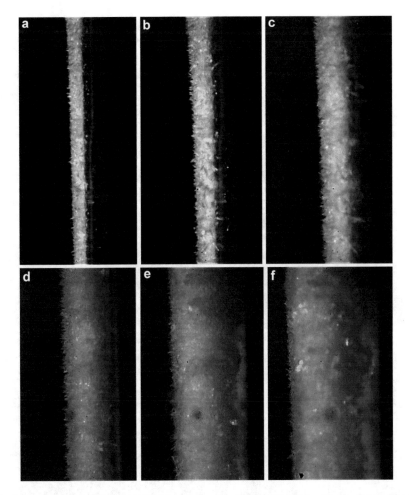

Fig. 6.31 Side view of the frost formation at test surface (TestID-2). (**a**) $t = 10$ min; (**b**) $t = 20$ min; (**c**) $t = 30$ min; (**d**) $t = 1$ h; (**e**) $t = 3$ h; (**f**) $t = 6$ h

Fig. 6.39. The temperature at the test surface increases with time as shown in Figs. 6.40 and 6.41. The temperature on the plain surface maintains at the melting temperature for a few minutes which is different than the strict rising on the superhydrophobic surface. The explanation might be that the frost layer attaches to the test surface on a plain surface but detaches from the test surface on a superhydrophobic surface during melting. The temperature variation inside the chamber is not significant. The temperature remains around 19.6 °C at the upper and 18.4 °C at the lower of the chamber for a test run, as shown in Fig. 6.42.

Defrost time and efficiency depend on frost mass, porosity, heating power, and surface wettability. The defrost process shortens on higher heating power. Table 6.5 shows that the defrost time decreases from 109 to 85 s on TestID-1, from 119 to 97 s on TestID-2, and from 165 to 120 s on TestID-3 when the heating power setting

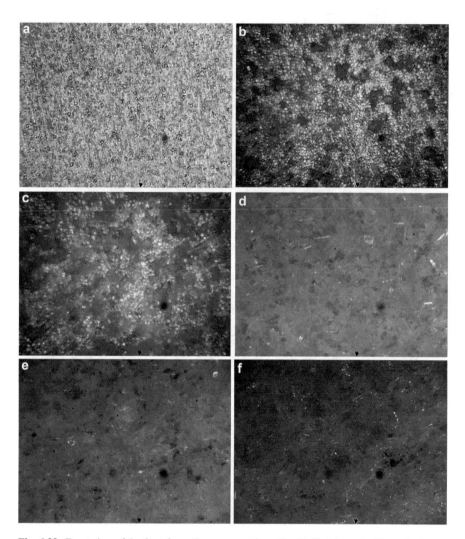

Fig. 6.32 Front view of the frost formation at test surface (TestID-3). (**a**) $t = 0$; (**b**) $t = 5$ min; (**c**) $t = 10$ min; (**d**) $t = 1$ h; (**e**) $t = 3$ h; (**f**) $t = 6$ h

increases from 15 to 60 W. However, the defrost efficiency is larger on lower heating power. Defrost efficiency on a wet surface increases from 12% to 41% on TestID-1 and 11% to 36% on TestID-2 when the heating power setting decreases from 60 to 15 W. The defrost efficiency on dry surface increases from 11% to 82% on TestID-3 when the heating power setting decreased from 60 to 6 W. The dependences of defrost time and defrost efficiency on heat flux are shown in Figs. 6.43 and 6.44. The defrost process was run on the superhydrophobic surface after frost grows for 4 h. The frost mass is 1.772, 1.561, and 1.8 g, corresponding to the average heat flux of 1226, 1642, and 3853 W/m². Among the three runs, the defrost time is the shortest

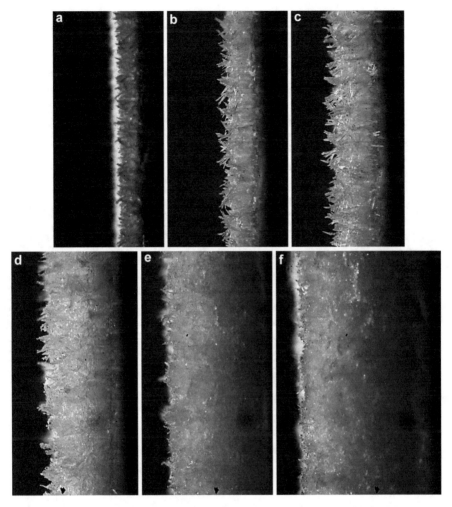

Fig. 6.33 Side view of the frost formation at test surface (TestID-3). (**a**) $t = 10$ min; (**b**) $t = 20$ min; (**c**) $t = 30$ min; (**d**) $t = 1$ h; (**e**) $t = 3$ h; (**f**) $t = 6$ h

with the highest average heat flux, and the defrost efficiency is the highest with the lowest average heat flux.

The effect of surface wettability on defrost time is shown in Figs. 6.45 and 6.46 for wet and dry surfaces. During defrost period, a thin film is retained on the superhydrophilic surface, and some nonuniform droplets stick on the plain surface. The test surface is wet when the frost melts away, but the water film or retention droplets are still on it. The defrost time is shorter on the superhydrophilic surface than that on the plain surface. The explanation is that the water film on the superhydrophilic surface improves the partial slumping of the frost and the drainage of the meltwater. The evaporation time is shorter on the superhydrophilic surface

Table 6.4 Frost properties

	Frost growth period (h)	Frost thickness (mm)	Frost mass (g)	Frost porosity
TestID-1	4	4.34	1.608	0.72
TestID-1	4	4.26	1.488	0.74
TestID-1	6	5.20	2.267	0.67
TestID-2	4	4.3	1.473	0.74
TestID-2	4	4.38	1.462	0.75
TestID-2	6	5.35	2.236	0.69
TestID-3	4	4.88	1.772	0.73
TestID-3	4	4.8	1.561	0.76
TestID-3	6	5.48	2.55	0.65
TestID-4	4	3.4544	2.78	0.65
TestID-4	4	3.4798	2.744	0.66
TestID-4	4	3.6322	2.617	0.69
TestID-5	4	3.71	2.979	0.65
TestID-5	4	3.86	2.966	0.67
TestID-5	8	5.2324	4.52	0.62
TestID-6	4	3.5814	2.761	0.66
TestID-6	4	3.5814	2.679	0.67
TestID-6	6	4.4196	4.08	0.6

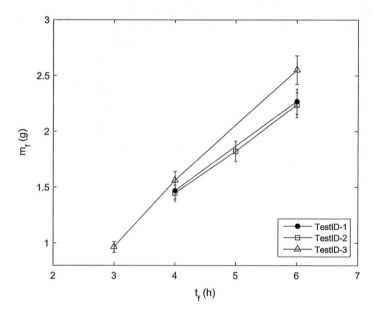

Fig. 6.34 Frost mass versus frost growth period

Fig. 6.35 Defrost process after frost grows for 4 h (TestID-1). (**a**) $t = 10$ s; (**b**) $t = 30$ s; (**c**) $t = 40$ s; (**d**) $t = 60$ s; (**e**) $t = 80$ s; (**f**) $t = 120$ s; (**g**) $t = 160$ s

than on the plain surface. The reason is that thin water film spreads over the surface and increases the contact area between the test surface and the water film. Heat transfer is thus larger on the superhydrophilic surface. The evaporation time is much longer on the plain surface as the contact area of water droplets to the test surface is much smaller. It is observed from the experiments that thin water film over the surface dries away quickly, but a thicker water film accumulates at the bottom of the test surface and takes quite some time to dry away. The evaporation time is much longer with lower heating power, compared to that with high heating power. The surface is dry when no water droplets or water film are on it. The defrost time is shorter on the superhydrophobic surface than on the superhydrophilic surface. When

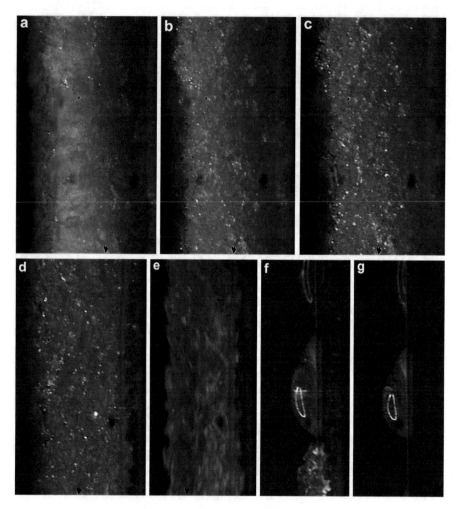

Fig. 6.36 Defrost process after frost grows for 4 h (TestID-2). (**a**) $t = 10$ s; (**b**) $t = 20$ s; (**c**) $t = 40$ s; (**d**) $t = 60$ s; (**e**) $t = 90$ s; (**f**) $t = 92$ s; (**g**) $t = 100$ s

frost melts on the superhydrophobic surface, the frost layer detaches from the test surface and falls off without retention of droplets. No evaporation time is required, and thus the defrost time is reduced significantly.

The influences of surface wettability on defrost efficiency are shown in Figs. 6.47 and 6.48. When the test plates are wet, the defrost efficiency is larger on the superhydrophilic surface than that on the plain surface due to the shorter defrost time. When the test plates are dry, the defrost efficiency is larger on the superhydrophobic surface compared to that on the superhydrophilic surface. The reason is that the defrost time is shorter on the superhydrophobic surface. The

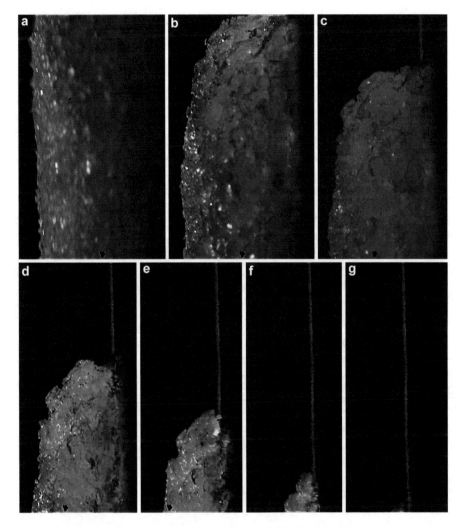

Fig. 6.37 Frost slumping process after frost grows for 6 h (TestID-3). (**a**) $t = 10$ s; (**b**) $t = 80$ s; (**c**) $t = 100$ s; (**d**) $t = 110$ s; (**e**) $t = 120$ s; (**f**) $t = 126$ s; (**g**) $t = 130$ s

improvement of defrost efficiency is noticeable with lower heating power. Defrost efficiency increases from 11% to 30% on the superhydrophobic surface compared to the increase from 7% to 16% on the superhydrophilic surface when the heating power setting decreases from 60 to 15 W. It can be explained by the increase of the evaporation time on the superhydrophilic surface with lower heating power.

A thin film heater is used to heat the large test plates (50 mm × 50 mm) with the actual power output of 7.7 W. The defrost properties are listed in Table 6.6. Compared to the heating by thermoelectric module, the average heat flux at the test plate is much smaller. The evaporation time is much longer due to the low heat

Table 6.5 Defrost properties for test surfaces 38 mm × 38 mm

	Frost mass (g)	Average heat flux applied (W/m^2)	Time to melting temperature (s)	Defrost time (s)	Time to dry out (s)	Defrost efficiency (dry surface)	Defrost efficiency (wet surface)
TestID-1	1.608	3525	25	89	171	0.07	0.13
TestID-1	1.488	1219	42	109	325	0.16	0.41
TestID-1	1.468	3394	26	85	185	0.06	0.12
TestID-2	1.473	4120	23	97	>240		0.11
TestID-2	1.422	1857	39	119	>320		0.36
TestID-2	1.516	1404	48	135	>400		0.88
TestID-3	1.800	3853	26	120		0.11	
TestID-3	1.561	1642	46	165		0.30	
TestID-3	1.772	1226	55	177		0.82	

"Dry surface" means that test surfaces are dried without water droplets or water film after heating
"Wet surface" means that test surfaces are still covered by water droplets on plain surface and by water film on superhydrophilic surface

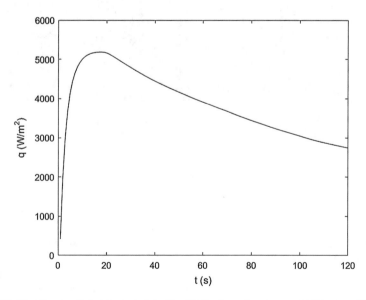

Fig. 6.38 Heat flux applied at the test plate (TestID-3) where direct current (DC) sets at 12 V/5 A

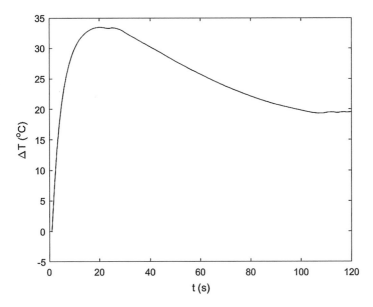

Fig. 6.39 Temperature difference between the heat flux sensor and the test plate (TestID-3) where DC sets at 12 V/5 A

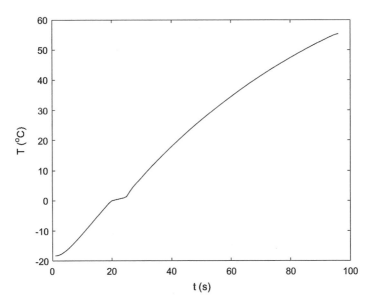

Fig. 6.40 Temperature at the test plate (TestID-2) during the defrost period where DC sets at 12 V/5 A

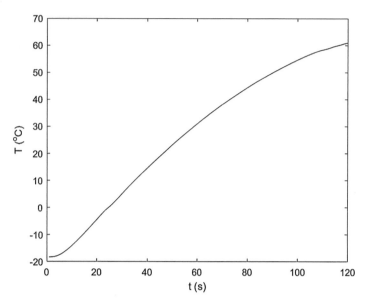

Fig. 6.41 Temperature at the test plate (TestID-3) during the defrost period where DC sets at 12 V/ 5 A

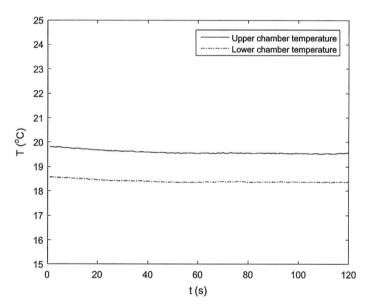

Fig. 6.42 Temperature inside the chamber during the defrost period where DC sets at 12 V/5 A

flux, and the defrost efficiency is higher than that with the high heating power. The dependence of defrost time and efficiency on surface wettability is shown in Figs. 6.49, 6.50, 6.51, and 6.52. When the test plates are wet, defrost time is shorter

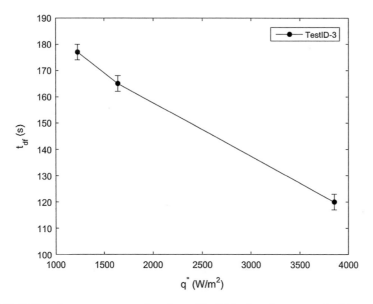

Fig. 6.43 Defrost time versus average heat flux at TestID-3 where frost grows for 4 h

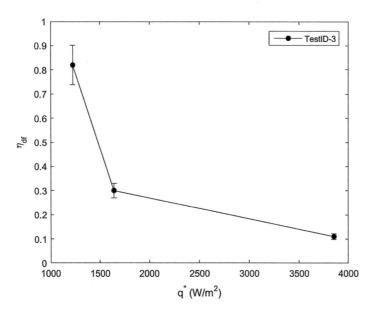

Fig. 6.44 Defrost efficiency versus average heat flux at TestID-3 where frost grows for 4 h

on the superhydrophilic surface compared to that on the plain surface. Defrost efficiency improves as well on the wet superhydrophilic surface. When the test plates are dry, defrost time is much shorter on the superhydrophobic surface than that on the superhydrophilic surface. It is observed that the evaporation time lasts

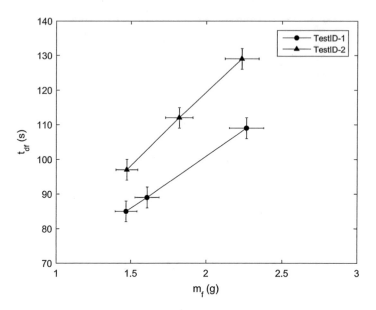

Fig. 6.45 Defrost time versus mass on wet surface where heating power sets at 60 W

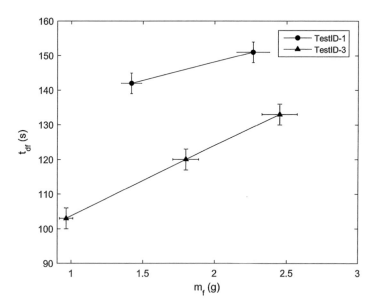

Fig. 6.46 Defrost time versus mass on dry surface where heating power sets at 60 W

over 600 s on the superhydrophilic surface and the plane surface. Defrost efficiency improves greatly on the superhydrophobic surface as there is no evaporation time for the defrost process.

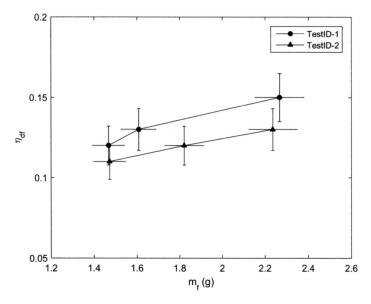

Fig. 6.47 Defrost efficiency versus mass on wet surface where heating power sets at 60 W

6.3 Comparison of Predictions and Measurements

Defrost processes are complicated because many factors affect the defrost mechanisms. Defrost time and efficiency are two key variables when controlling defrost processes in applications. The relationship between defrost time and efficiency and defrost mechanisms are discussed in the model and validated by the experiments. Surface wetting properties could influence frost properties as well as defrost processes.

A defrost process is divided into three stages based on motions of meltwater. The magnitude of volume flux of water, melting velocity and draining velocity determines the motions of meltwater as well as defrost mechanisms. When the volume flux of water is greater than melting velocity, the meltwater is absorbed into the frost layer, and absorption was validated at the early period during the defrost process. With water saturation increasing, volume flux of water decreases, and meltwater starts to drain. When the volumetric draining rate is less than the volumetric rate of melting water, part of the meltwater accumulates between the surface and saturated frost layer. Frost could fall off with the draining water. Draining was observed in the middle of the defrost process, and the remaining frost finally fell off.

The effects of surface wettability are discussed in the model and the slumping criterion. Surface wetting properties could influence the volume flux of water and melting velocity by affecting the frost properties. The average volumetric draining rate is predicted to increase by 7.1% on hydrophobic surfaces due to the slip velocity at the surface. The draining velocity is not measured in the experiments. However,

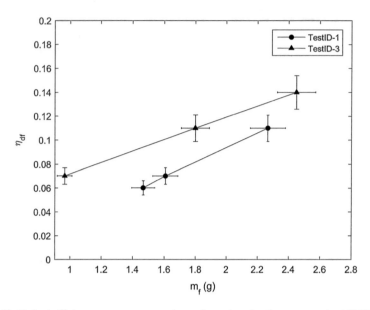

Fig. 6.48 Defrost efficiency versus mass on dry surface where heating power sets at 60 W

Table 6.6 Defrost properties for test surfaces 50 mm × 50 mm

	Frost mass (g)	Average heat flux applied (W/m^2)	Time to melting temperature (s)	Defrost time (s)	Time to dry out (s)	Defrost efficiency (dry surface)	Defrost efficiency (wet surface)
TestID-4	2.78	417	62	165	>600	<0.22	0.8
TestID-4	2.744	339	62	158	>600	<0.22	0.83
TestID-4	2.86	301	61	153	>600	<0.23	0.9
TestID-5	2.577	407	64	167	>600		0.73
TestID-5	2.828	355	69	205	>600		0.66
TestID-5	2.966	336	65	178	>600		0.8
TestID-6	2.679	247	68	177		0.71	
TestID-6	2.711	230	74	161		0.79	
TestID-6	2.992	190	72	184		0.76	

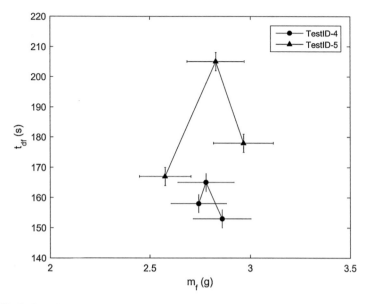

Fig. 6.49 Defrost time versus mass on wet surface where actual heating power is 7.7 W

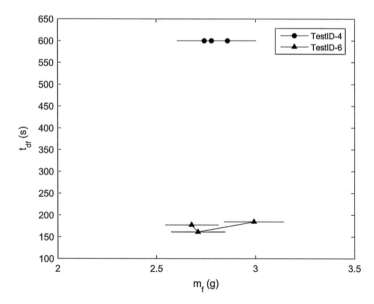

Fig. 6.50 Defrost time versus mass on dry surface where actual heating power is 7.7 W

the prediction agrees with the literature [41] in which the draining velocity is shown to be larger on hydrophobic surfaces than that on hydrophilic surfaces. The slumping force ratio is greater on hydrophobic surfaces as predicted in the model. In

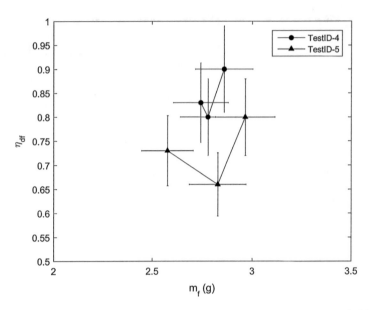

Fig. 6.51 Defrost efficiency versus mass on wet surface where actual heating power is 7.7 W

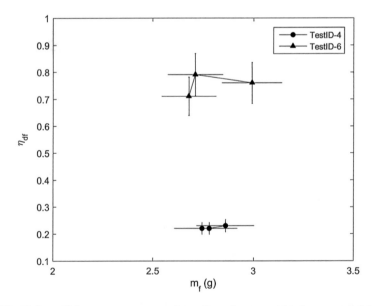

Fig. 6.52 Defrost efficiency versus mass on dry surface where actual heating power is 7.7 W

experiments, frost slumping occurred on superhydrophobic surfaces during the defrost processes.

Defrost time and efficiency depend on the defrost mechanisms. The defrost model predicts the defrost time larger compared to the values measured in the experiments.

The reason is that frost slumping reduced the defrost time. The model could better predict defrost time when coupling with the frost slumping criterion. In applications, geometry and other related factors have to be taken into consideration when formulating frost slumping criteria.

Chapter 7
Conclusion

This research reported in this monograph is motivated by the current problem in refrigeration system that how would surface wettability influences the defrost process. The defrost process is analytically formulated based on its physical mechanisms. In the literature, meltwater is either assumed to be absorbed into the frost layer or drained in the current defrost models. However, the factors affecting the meltwater behavior are not studied. In this work, water permeation rate is inspected in absorption, and drainage rate is investigated with respect to the different surface wettability. The slumping criterion is studied as a potential method of frost removal. Slumping conditions vary on surfaces with different surface wettability. Superhydrophilic and superhydrophobic surfaces are prepared and examined in the experiments, and the results show that frost falls off as rigid pieces on the superhydrophobic surface while it melts with partial slumping on the superhydrophilic surface.

A general defrost process consists of surface preheating, frost melting, and evaporation of retention water. This work is focused on frost melting and meltwater behaviors. Three variables determine the melting scheme. They are melting rate, water permeation rate, and draining rate. Melting rate depends on the heat flux applied at the surface and the porosity of the frost layer. Water permeation rate is a function of water saturation and frost layer permeability. Draining rate is related to the velocity boundary conditions at the interfaces. The competition among the three variables determines whether meltwater is absorbed into the frost layer or drains along the test surface. Based on these studies, the melting process is divided into three stages: absorption, accumulation, and drainage. In the absorption stage, meltwater is absorbed into the frost layer by capillary force. Water saturation is formulated with mass continuity and is shown as a function of time and location. A thin water film could form over the test surface due to surface tension around the perimeters of the frost column in accumulation. In the drainage stage, meltwater drains by gravity. The draining rate varies with the velocity at the solid–water interface and at the water–permeation interface.

© The Author(s), under exclusive license to Springer Nature Switzerland AG 2019   105
Y. Liu, F. A. Kulacki, *The Effect of Surface Wettability on the Defrost Process*,
SpringerBriefs in Applied Sciences and Technology,
https://doi.org/10.1007/978-3-030-02616-5_7

Frost slumping is a phenomenon in which frost columns fall off from the surface as rigid pieces or as parts of the bulk mass. In comparison with the traditional defrost methods, frost removal by slumping could reduce defrost time and energy consumption. In this work, a slumping criterion is formulated based on the static force analysis. A slumping force ratio is defined as the ratio of gravity to adhesive force at the surface. Both gravity and adhesive force change with defrost time. In the absorption/accumulation stage, ice adhesion/surface tension attaches the frost mass to the surface, and the strength of ice adhesion/surface tension decreases with increases of the surface temperature. In the drainage stage, both the gravity and surface tension decrease, and the force ratio depends on the drainage rate of the meltwater and the surface temperature.

Surface wettability influences the defrost process and the slumping criterion. In the absorption stage, water permeation rate is related to the structure of the frost layer, which varies with surface wettability. The water flux increases with the permeability of the frost layer, which increases with the frost porosity. The draining rate is a function of the velocity at the solid–water interface. A no slip boundary condition is assumed on the hydrophilic surface, and a slip boundary condition is applied to the hydrophobic surface. The model shows that the draining rate increases about 7% on hydrophobic surfaces compared to that on hydrophilic surfaces. Slumping criterion is formulated for hydrophilic surface and hydrophobic surface separately in the accumulation and drainage stages. The expressions of surface tension are different for hydrophilic surfaces and for hydrophobic surfaces. A thin water film is assumed on hydrophilic surfaces, and the surface tension is a function of the contact angle. Water droplets are assumed on hydrophobic surfaces, and the surface tension is a function of advancing contact angle and receding contact angle.

Experiments have been conducted on a superhydrophilic surface, plain surface, and superhydrophobic surface. The defrost process is observed to be different on the three types of surfaces. On the superhydrophilic and plain surfaces, frost melts at the center of the surface very quickly when the surface temperature reaches the melting temperature. The meltwater divides into two parts. One part is absorbed into the thicker frost layer at the edges, and the other part retains on the surface in the form of water droplets on plain surfaces and water film on the superhydrophilic surface. On the edges of the test plate, the frost at the bottom melts and drains first. The frost at the top falls with draining water after a period of time. On the superhydrophobic surface, the center area of the frost column seems to detach from the surface when the surface temperature reaches the melting temperature. The frost column still adheres to the surface when the frost at the edges attaches to the surface and finally falls off as a rigid whole piece. After the defrost process, no water retention is observed on the superhydrophobic surface, while water droplets are retained on the plain surface and water film is retained on the superhydrophilic surface. When the evaporation time of water droplets or water film is not considered, defrost time and efficiency do not vary significantly on the three types of surfaces. However, defrost time and efficiency are improved considerably when the water droplets or water film are evaporated. Heating powers could influence defrost time and efficiency during the defrost process. With higher heating power, the defrost time is shorter, but the defrost efficiency is lower. With lower heating power, the defrost time is longer, but the defrost efficiency is longer. Defrost time and efficiency also depend on ambient temperature, system design, and defrost methods.

7.1   Future Research Pathways

Current work includes formulating the mathematical model of the melting process, analyzing the influence of the surface wettability on the melting process quantitatively and theoretically, and investigating the slumping criterion during the defrost process. The research could be explored and extended in many aspects.

The formulation of the mass continuity in the permeation layer is one dimensional based on the assumptions that the frost layer is uniform. A multidimensional formulation could better describe the frost layer with nonuniform properties. The frost layer melts with a moving boundary. The thickness of the frost layer decreases with time. The analytical solution might be possible in the future study. The boundary conditions are assumed to be fixed at the solid–permeation interface and at the permeation–frost interface. In the application, the practical boundary conditions should be included, and surface wettability might influence the frost structures and thus the permeation rate.

The formula of the permeation flux can be modified to accommodate the frost column. In the formula, the intrinsic permeability, the entry capillary pressure, and the power law constants are not available in the current literature. The values in the references are used in this work. In the application, the values should be determined by experiments. Visualization scheme might be applied to track the saturation ratio and the permeation front with time.

Calibration of the convection heat transfer coefficient can be improved. The heat transfer coefficient at the test surface–air interface is calculated by measuring the temperature at the test surface and the temperature of air that is very close to the test surface. A heat flux sensor is attached to the test surface to obtain the heat flux across the surface. The calculated value is small compared to that from the correlations. The error might result from the inaccuracy of the air temperature measurement.

The thermal boundary condition at the test surface is expected to be controllable. In current design, a thermal electric module is applied for cooling cycle, and a film heater is used for heating cycle. During the defrost cycle, the test plate experiences a transient state when heat is applied at the plate. Heat flux across the test surface would not be constant. In refrigeration industries, hot gas purge might be applied in the tube for the defrost process, so the constant temperature boundary condition can be considered when setting up the model. The temperature boundary condition can be realized by remodeling the current design with a constant temperature controller and a bath. The coolant absorbs the heat from the test plate quickly, and the temperature of the coolant increases significantly in the current system.

The test plate is placed inside the chamber without fastening to avoid the interference with the frost and defrost process. Thermal resistance between the test plate and the cooling/heating apparatus thus increases. A mechanical fixture could help reduce the thermal resistance if the design does not affect the frost and defrost process. The measurement of frost properties could be improved. The frost growth period is accompanied by the thickening of the frost density and the growing of the frost thickness. Frost porosity varies with time and location. In current measurement, a uniform density is assumed. Also, the frost layer is denser and thicker at the edges. In the future, a round geometry of the test plate might be used to avoid the edge effects.

The range of the parameters is restricted by the cooling method of the chamber. The chamber is cooled by thermal electric modules, and heat is removed by the heat sink extended inside the chamber. The sink temperature becomes very low in order to achieve low temperature inside the chamber. Frost might form on the heat sink when the relative humidity in the chamber is high. In the future, the air could be cooled down from a different system, and cooling air could be circulated through the test chamber.

The physical mechanism of slumping is more complicated with the involvement of the physics of interfacial forces. Herein the slumping condition is based on the force balance analysis. Further research can formulate the slumping criterion with respect to the physical mechanism of fracture or crack of the frost column. In that case, the cohesive forces of the frost and the adhesive forces at the interfaces should be considered.

The slumping criterion is expected to be verified on surfaces with different contact angles and contact angle hysteresis. In the current experiments, surface treatment is limited to a superhydrophilic surface and a superhydrophobic surface. The contact angle hysteresis is not controllable for the hydrophobic surface. The slumping criterion on hydrophobic surface depends on the advancing/receding contact angles. The rolling of the droplet occurs with small contact angle hysteresis and proper droplet volume and sliding angle. In the experiments, surface treatment methods also lead to different defrost mechanisms. Frost might fall off as a rigid body or break up into fractures on the different superhydrophobic surfaces. More samples with different surface properties are expected in the experiments.

Tests on reliability and repeatability of the surface effects are expected. In the current experiments, the surface shows good superhydrophobicity, and the results are relatively consistent. Frost slumping occurs after different frost formation periods and with different defrost heating methods. In rare cases, it is observed that few small droplets stick to the superhydrophobic surface after the defrost process. The droplets do not roll off after the frost and defrost cycle.

The experiments are expected to test with large scale equipment. In the application, coils have different geometries. In the current study, frost and defrost cycles are on a flat aluminum plate placed vertically. A complicated multidimensional model can be formulated with consideration of fin arrays and interaction between fins.

Instruments and Precision

Measurement variables	Instruments	Model	Range	Accuracy
Chamber temperature	Relative humidity/temperature transmitter	Omega Engineering: HX94CW	0–100 °C	±0.6 °C
Chamber temperature	Quick disconnection TCs	Omega Engineering: SCPSS-062G-6		
Chamber humidity	Relative humidity/temperature transmitter	Omega Engineering: HX94CW	3–95%	±2%
Test plate temperature	PFA insulated T/Cs	Omega Engineering: 5TC-TT-T-30-72	−267 to 260 °C or −450 to 500 F	±0.5 °C
Heat flux	Thin-film heat flux sensor	Omega Engineering: HFS-4	<30,000 Btu/Ft^2-Hr	Sensitivity of 6.5 μV/Btu/Ft^2-Hr; ±0.5%
Heat flux temperature	Thin-film heat flux sensor with built-in T/C Type K	Omega Engineering: HFS-4	−200 to 150 °C or −330 to 300 F	
Mass	Digital scale	Acculab: ALC-320.3	0–320 g	±3 mg
Frost image	High-intensity illuminator	Edmund Fiber-Lite Optics: MI-150		
Frost front/side profile	5.1 megapixel CMOS color camera	BigCatch: DCMC510		

(continued)

© The Author(s), under exclusive license to Springer Nature Switzerland AG 2019
Y. Liu, F. A. Kulacki, *The Effect of Surface Wettability on the Defrost Process*,
SpringerBriefs in Applied Sciences and Technology,
https://doi.org/10.1007/978-3-030-02616-5

Measurement variables	Instruments	Model	Range	Accuracy
Frost front profile	Zoom imaging lens	Edmund Optics: VZM 450i, 4.5×		
Frost side profile	Zoom imaging lens	Navitar Zoom 7000, 6.0×, 6.4 mm; 0.5″ CCD mono-chrome sensor	25 frame/s	

Slumping Image

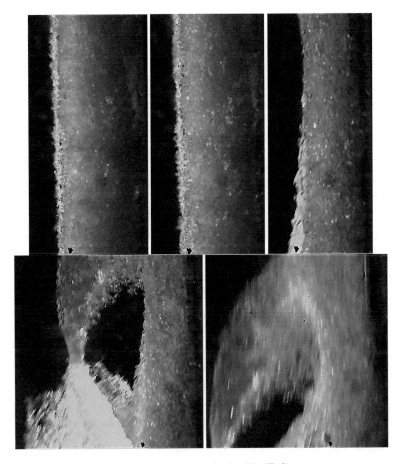

Fig. B.1 Frost slumping process after frost grows for 4 h (TestID-3)

Y. Liu, F. A. Kulacki, *The Effect of Surface Wettability on the Defrost Process*,
SpringerBriefs in Applied Sciences and Technology,
https://doi.org/10.1007/978-3-030-02616-5

References

1. Na B, Webb RL (2003) A fundamental understanding of factors affecting frost nucleation. Int J Heat Mass Transfer 46:3797–3808
2. Piucco RO, Hermes CJL, Melo C, Barbosa JR Jr (2008) A study of frost nucleation on flat surfaces. Exp Therm Fluid Sci 32:1710–1715
3. Thibaut Brian PL, Reid RC, Shah YT (1970) Frost deposition on cold surfaces. Ind Eng Chem Fund 9(3):375–380
4. Kennedy LA, Goodman J (1974) Free convection heat and mass transfer under conditions of frost deposition. Int J Heat Mass Transfer 17:477–484
5. Hayashi Y, Aoki A, Adachi S, Hori K (1977) Study of frost properties correlating with frost formation types. Trans ASME J Heat Transfer 99:239–245
6. Cremers CJ, Mehra VK (1982) Frost formation on vertical cylinders in free convection. Trans ASME J Heat Transfer 104:3–7
7. Fossa M, Tanda G (2002) Study of free convection frost formation on a vertical plate. Exp Therm Fluid Sci 26:661–668
8. Liu FZ, Chen HX, Fu JK (2002) Study on frost characteristics of finned-tube heat exchanger under low temperature conditions. Fluid Machinery 30(11):54–57
9. Wu X, Dai W, Xu W, Tang L (2007) Mesoscale investigation of frost formation on a cold surface. Exp Therm Fluid Sci 31:1043–1048
10. Janssen DD, Mohs WF, Kulacki FA (2016) Frost layer growth based on high-resolution image analysis. Trans ASME 8:021018-1–021018-12
11. Jones BW, Parker JD (1975) Frost formation with varying environmental parameters. Trans ASME J Heat Transfer 97(2):255–259
12. Schneider HW (1978) Equation of the growth rate of frost forming on cooled surfaces. Int J Heat Mass Transfer 21:1019–1024
13. Dietenberger MA (1983) Generalized correlation of the water frost thermal conductivity. Int J Heat Mass Transfer 26(4):607–619
14. Tao Y-X, Besant RW, Rezkallah KS (1993) A mathematical model for predicting the densification and growth of frost on a flat plate. Int J Heat Mass Transfer 36(2):353–363
15. Tao Y-X, Besant RW (1993) Prediction of spatial and temporal distributions of frost growth on a flat plate under forced convection. Trans ASME J Heat Transfer 115:278–281
16. Lee K-S, Kim W-S, Lee T-H (1997) A one-dimensional model for frost formation on a cold flat surface. Int J Heat Mass Transfer 40(18):4359–4365
17. Le Gall R, Grillot JM (1997) Modelling of frost growth and densification. Int J Heat Mass Transfer 40(13):3177–3187

18. Cheng C-H, Cheng Y-C (2001) Predictions of frost growth on a cold plate in atmospheric air. Int Commun Heat Mass Transfer 28(7):953–962
19. Lee K-S, Jhee S, Yang D-K (2003) Prediction of the frost formation on a cold flat surface. Int J Heat Mass Transfer 46:3789–3796
20. Na B, Webb RL (2004) Mass transfer on and within a frost layer. Int J Heat Mass Transfer 47:899–911
21. Lee YB, Ro ST (2005) Analysis of the frost growth on a flat plate by simple models of saturation and Supersaturation. Exp Therm Fluid Sci 29:685–696
22. Na B, Webb RL (2004) New model for frost growth rate. Int J Heat Mass Transfer 47:925–936
23. Hao YL, Iragorry J, Tao Y-X (2005) Frost-air interface characterization under natural convection. Trans ASME J Heat Transfer 127:1174–1180
24. Lenic K, Trp A, Frankovic B (2006) Unsteady heat and mass transfer during frost formation in a fin-and-tube heat exchanger. Energy Environ:35–48
25. Sahin AZ (1995) An analytical study of frost nucleation and growth during the crystal growth period. Heat Mass Transfer 30:321–330
26. Sahin AZ (2000) Effective thermal conductivity of frost during the crystal growth period. Int J Heat Mass Transfer 43:539–553
27. Shin J, Tikhonov AV, Kim C (2003) Experimental study on frost structure on surfaces with different hydrophilicity: density and thermal conductivity. Trans ASME J Heat Transfer 125 (1):84–94
28. Zhong YF, Jacobi AM, Georgiadis JG (2006) Condensation and wetting behavior on surfaces with micro-structures: super-hydrophobic and super-hydrophilic. Proc Int Ref Air Cond, Paper 828
29. Liu ZL, Wang HY, Zhang XH, Meng S, Ma CF (2006) An experimental study on minimizing frost deposition on a cold surface under natural convection conditions by use of a novel anti-frosting paint, Part I. Int J Refrigeration 29:229–236
30. Liu ZL, Zhang XH, Wang HY, Meng S, Cheng S (2007) Influences of surface hydrophilicity on frost formation on a vertical cold plate under natural convection conditions. Exp Therm Fluid Sci 31(7):789–794
31. Liu ZL, Gou YJ, Wang JT, Cheng S (2008) Frost formation on a super-hydrophobic surface under natural convection conditions. Int J Heat Mass Transfer 51(25–26):5975–5982
32. Chen C-H, Cai Q, Tsai C, Chen C-L, Xiong G, Yu Y, Ren Z (2007) Dropwise condensation on superhydrophobic surfaces with two-tier roughness. Appl Phys Lett 90:173108
33. Wang H, Tang LM, Wu XM, Dai WT, Qiu YP (2007) Fabrication and anti-frosting performance of superhydrophobic coating based on modified nano-sized calcium carbonate and ordinary polyacrylate. Appl Surf Sci 253(22):8818–8824
34. Wang FC, Li CR, Lv YZ, Du YF (2009) A facile superhydrophobic surface for mitigating ice accretion. In: Proceedings of the 9th international conference on properties and applications of dielectric materials A-34:150–153
35. Varanasi KP, Deng T, Smith JD, Hsu M, Nitin Bhate N (2010) Frost formation and ice adhesion on superhydrophobic surfaces. Appl Phys Lett 97(23):234102
36. He M, Wang JX, Li HL, Jin XL, Wang JJ, Liu BQ, Song YL (2010) Super-hydrophobic film retards frost formation. Soft Matter 6:2396–2399
37. He M, Wang JX, Li HL, Song YL (2011) Super-hydrophobic surfaces to condensed micro-droplets at temperatures below the freezing point retard ice/frost formation. Soft Matter 7:3993–4000
38. Farhadi S, Farzaneh M, SKulinich SA (2011) Anti-icing performance of superhydrophobic surfaces. Appl Surf Sci 257(14):6264–6269
39. Bahadur V, Mishchenko L, Hatton B, Taylor JA, Aizenberg J, Krupenkin T (2011) Predictive model for ice formation on superhydrophobic surfaces. Langmuir 27(23):14143–14150
40. Min J, Webb RL, Bemisderfer CH (2000) Long-term hydraulic performance of dehumidifying heat-exchangers with and without hydrophilic coatings. HVAC&R Res 6(3):257–272

41. Jhee S, Lee K-S, Kim W-S (2002) Effect of surface treatments on the frosting/defrosting behavior of a fin-tube heat exchanger. Int J Refrigeration 25:1047–1053

42. Kim K, Lee KS (2011) Frosting and defrosting characteristics of a fin according to surface contact angle. Int J Heat Mass Transfer 54(13–14):2758–2764

43. Wu XM, Webb RL (2001) Investigation of the possibility of frost release from a cold surface. Exp Therm Fluid Sci 2(3–4):151–156

44. Huang LY, Liu ZL, Liu YM, Gou YJ, Wang JT (2009) Experimental study on frost release on fin-and-tube heat exchangers by use of a novel anti-frosting paint. Exp Therm Fluid Sci 33:1049–1054

45. Antonini C, Innocenti M, Horn T, Marengo M, Amirfazli A (2011) Understanding the effect of superhydrophobic coatings on energy reduction in anti-icing systems. Cold Reg Sci Technol 67:58–67

46. Jing T, Kim Y, Lee S, Kim D, Kim J, Hwang W (2013) Frosting and defrosting on rigid superhydrophobic surface. Appl Surf Sci 276:37–42

47. Boreyko JB, Srijanto BR, Nguyen TD, Carlos Vega C, Fuentes-Cabrera M, Collier CP (2013) Dynamic defrosting on nanostructured superhydrophobic surfaces. Langmuir 29:9516–9524

48. Chen XM, Ma RY, Zhou HB, Zhou XF, Che LF, Yao SH, Wang ZK (2013) Activating the microscale edge effect in a hierarchical surface for frosting suppression and defrosting promotion. Sci Rep 3:2515

49. Korte C, Jacobi AM (2001) Condensate retention effects on the performance of plain-fin-and-tube heat exchangers: retention data and modeling. Trans ASME J Heat Transfer 123(5):926–936

50. Min J, Webb RL (2001) Condensate formation and drainage on typical fin materials. Exp Therm Fluid Sci 25:101–111

51. Zhong Y, Joardar A, Gu Z, Park Y-G, Jacobi AM (2005) Dynamic dip testing as a method to assess the condensate drainage behavior from the air-side surface of compact heat exchangers. Exp Therm Fluid Sci 29:957–970

52. El Sherbini AI, Jacobi AM (2006) A model for condensate retention on plain-fin heat exchangers. Trans ASME J Heat Transfer 128:427–433

53. Sommers AD, Jacobi AM (2008) Wetting phenomena on micro-grooved aluminum surfaces and modeling of the critical droplet size. J Colloid Interface Sci 328(2):402–411

54. Liu L, Jacobi AM (2009) Air-side surface wettability effects on the performance of slit-fin-and-tube heat exchangers operating under wet-surface conditions. Trans ASME J Heat Transfer 13:051802-1–051802-9

55. Rahman AM, Jacobi AM (2012) Drainage of frost meltwater from vertical brass surfaces with parallel microgrooves. Int J Heat Mass Transfer 55:1596–1605

56. Sanders CT (1974) The influence of frost formation and defrosting on the performance of air coolers. Doctoral dissertation, Delft University of Technology

57. Krakow KI, Yan L, Lin S (1992) A model of hot-gas defrosting of evaporators. Part 1: Heat and mass transfer theory. ASHRAE Trans 98(1):451–461

58. Krakow KI, Yan L, Lin S (1992) A model of hot-gas defrosting of evaporators. Part 2: Experimental analysis. ASHRAE Trans 98(1):462–474

59. Sherif SA, Hertz MG (1998) A semi-empirical model for electric defrosting of a cylindrical coil cooler. Int J Energy Res 22(1):85–92

60. Lamberg P, Siren K (2003) Analytical model for melting in a semi-infinite PCM storage with an internal fin. Heat Mass Transfer 39:167–176

61. Liu Z, Tang G, Zhao F (2003) Dynamic simulation of air-source heat pump during hot-gas defrost. Appl Therm Eng 23:675–685

62. Hoffenbecker N, Klein SA, Reindl DT (2005) Hot gas defrost model development and validation. Int J Refrigeration 28(4):605–615

63. Dopazo JA, Fernandez-Seara J, Uhia FJ, Diz R (2010) Modelling and experimental validation of the hot-gas defrost process of an air-cooled evaporator. Int J Refrigeration 33(4):829–839

64. Minglu Q, Liang X, Shiming D, Yiqiang J (2012) A study of the reverse cycle defrosting performance on a multi-circuit outdoor coil unit in an air source heat pump. Part I: Experiments. Appl Energy 91:122–129

65. Qu M, Pan D, Xia L, Deng S, Jiang Y (2012) A study of the reverse cycle defrosting performance on a multi-circuit outdoor coil unit in an air source heat pump. Part II: Modeling analysis. Appl Energy 91:274–280

66. Mohs WF (2012) Heat and mass transfer during the melting process of a porous frost layer on a vertical surface. Doctoral dissertation, University of Minnesota

67. Raraty LE, Tabor D (1958) The adhesion and strength properties of ice. Proc R Soc Lond A 245:84–201

68. Jellinek HHG (1959) Adhesive properties of ice. J Colloid Interface Sci 14:268–280

69. Ryzhkin IA, Petrenko VF (1997) Physical mechanisms responsible for ice adhesion. J Phys Chem B 101(32):6267–6270

70. Makkonen L (2012) Ice adhesion – theory, measurements and countermeasures. J Adhes Sci Technol 26:413–445

71. Chen J, Liu J, He M, Li K, Cui D, Zhang Q, Zeng X, Zhang Y, Wang J, Song Y (2012) Superhydrophobic surfaces cannot reduce ice adhesion. Appl Phys Lett 101(11):111603-1–111603-3

72. Meuler AJ, Smith JD, Varanasi KK, Mabry JM, McKinley GH, Cohen RE (2010) Relationships between water wettability and ice adhesion. ACS Appl Mater Interfaces 2 (11):3100–3110. https://doi.org/10.1021/am1006035

73. Majumdar A, Mezic I (1999) Instability of ultra-thin water films and the mechanism of droplet formation on hydrophilic surfaces. Trans ASME J Heat Transfer 121:964–971

74. Aoki K, Hattori M, Ujiie T (1988) Snow melting by heating from the bottom surface. JSME Int J 31(2):269–275

75. Colbeck SC, Davidson G (1972) Water percolation through homogeneous snow. IASH Publication 107:242–257

76. Colbeck SC (1974) The capillary effects on water percolation in homogeneous snow. J Glaciol 13(67):85–97

77. Colbeck SC (1976) An analysis of water flow in dry snow. Water Resour Res 12(3):523–527

78. Colbeck SC (1982) The permeability of a melting snow cover. Water Resour Res 18 (4):904–908

79. Bengtsson L (1982) Percolation of meltwater through a snowpack. Cold Reg Sci Technol 6:73–81

80. Whitaker S (1986) Flow in porous media I: a theoretical derivation of Darcy's law. Transport Porous Med 1:3–25

81. Shaun Sellers S (2000) Theory of water transport in melting snow with moving surface. Cold Reg Sci Technol 31:47–57

82. Manthey S, Hassanizadeh SM, Helmig R, Hilfer R (2008) Dimensional analysis of two-phase flow including a rate-dependent capillary pressure-saturation relationship. Adv Water Resour 31:1137–1150

83. Daanen RP, Nieber JL (2009) Model for coupled liquid water flow and heat transport with phase change in a snowpack. J Cold Reg Eng 23(2):43–68

84. Hirashima H, Yamaguchi S, Sato A, Lehning M (2010) Numerical modeling of liquid water movement through layered snow based on new measurements of the water retention curve. Cold Reg Sci Technol 64:94–103

85. Yamaguchi S, Katsushima T, Sato A, Kumakura T (2010) Water retention curve of snow with different grain sizes. Cold Reg Sci Technol 64:87–93

86. Szymkiewicz A (2013) Modeling water flow in unsaturated porous media. Springer-Verlag, Berlin, Heidelberg

87. Katsushima T, Satoru Yamaguchi S, Kumakura T, Atsushi Sato A (2013) Experimental analysis of preferential flow in dry snowpack. Cold Reg Sci Technol 85:206–216

88. Washburn EW (1921) The dynamics of capillary flow. Phys Rev 18(3):273–283

89. Tsypkin GG (2010) Effect of the capillary forces on the moisture saturation distribution during the thawing of a frozen soil. Fluid Dyn 45(6):942–951
90. Beavers GS, Joseph DD (1967) Boundary conditions at a naturally permeable wall. J Fluid Mech 30(1):197–207
91. Taylor GI (1971) A model for the boundary condition of a porous material. Part 1. J Fluid Mech 49(2):319–326
92. Richardson S (1971) A model for the boundary condition of a porous material. Part 2. J Fluid Mech 49(2):327–336
93. Sahraoui M, Kaviany M (1992) Slip and no-slip velocity boundary conditions at interface of porous, plain media. Int J Heat Mass Transfer 35(4):927–943
94. Vinogradova OI (1995) Drainage of a thin liquid film confined between hydrophobic surfaces. Langmuir 11(6):2213–2220
95. Barrat JL (1999) Large slip effect at a nonwetting fluid-solid interface. Phys Rev Lett 82 (23):4671–4674
96. Baidry J, Charlaix E (2001) Experimental evidence for a large slip effect at a nonwetting fluid-solid interface. Langmuir 17(17):5232–5236
97. de Gennes PG (2002) On fluid/wall slippage. Langmuir 18(9):3413–3414
98. Andrienko D, Dünweg B (2003) Boundary slip as a result of a prewetting transition. J Chem Phys 119(24):13106–13112
99. Lauga E, Brenner MP, Stone HA (2005) Chapter 15: Microfluidics: the no-slip boundary condition. In: Foss J, Tropes C, Yarin A (eds) Handbook of experimental fluid dynamics. Springer, New York
100. Choi C-H, Kim C-J (2006) Large slip of aqueous liquid flow over a nanoengineered superhydrophobic surface. Phys Rev Lett 96:066001
101. Crank J (1984) Free and moving boundary problems. Oxford University Press, New York
102. Alexiades V, Solomon AD (1993) Mathematical modeling of melting and freezing. Hemisphere, Washington
103. Morton KW, Mayers DF (2005) Numerical solutions of partial differential equations. In: Cambridge University Press. Cambridge, England
104. Hamming RW (1973) Numerical methods for scientists and engineers. Dover Publications, Inc., New York
105. Kahraman R, Zughbi HD, Al-Nassar N (1998) A numerical simulation of melting of ice heated from above. Math Comput Appl 3(3):127–137
106. Lee TE, Baines MJ, Langdon S (2015) A finite difference moving mesh method based on conservation for moving boundary problems. J Comput Appl Math 288:1–17
107. Liu Y (2017) Effect of surface wettability on the defrost process. Doctoral dissertation, University of Minnesota, Minneapolis

Printed in the United States
By Bookmasters